"No one working in historical animal studies today surpasses Helen Cowie's depth and breadth of knowledge. A keen analyst of the cultural and economic demands humans place on other animals, Cowie never writes of non-human animals as mere resources. Using historical documents, archaeology and the biological sciences, her accounts give voice to animal sentience and agency. From turnspit dogs who refused to work on their day off to an alpaca recorded as plaintively moaning in remembrance of two conspecifics who died during their nineteenth-century voyage from Peru to Australia, Cowie's animals are not Cartesian stimulus-response machines. *Animals in World History* is the summary overview needed to move a growing field forward."

Abel A. Alves, *Professor of History, Ball State University*

W0018554

Animals in World History

This volume provides a concise synthesis of human–animal relations over time, charting shifting attitudes towards animals from domestication to the present day. It asks how non-human species have shaped human history, and how humans have reconfigured the animal world.

Humans have had a long and close relationship with animals. They have hunted them, consumed them as food and fashion, exploited them as energy sources, utilised them in warfare, exhibited them in zoos and menageries, and studied them for science. In the process, they have radically changed the way in which many animals live, subjecting them to captivity, altering their diets, constraining their movements and, through selective breeding, reshaping their bodies. The book explores the use of animals for sustenance, labour, companionship and display, and traces the rise of the animal rights movement. It also assesses how humans have impacted the overall biodiversity of the planet, driving some species of animals to extinction and permitting others to colonise new continents.

With case studies on animal astronauts, celebrity kakapos, globetrotting pandas and cocaine hippos, *Animals in World History* offers a lively and accessible introduction to human–animal relations for students and instructors of animal studies, environmental history, and social and cultural history.

Helen Louise Cowie is Professor of History at the University of York. She is the author of *Conquering Nature in Spain and its Empire, 1750–1850* (2011), *Exhibiting Animals in Nineteenth-Century Britain: Empathy, Education, Entertainment* (2014), *Llama* (2017) and *Victims of Fashion: Animal Commodities in Victorian Britain* (2021).

Themes in World History
Series editors: Peter N. Stearns and Jane Hooper

The *Themes in World History* series offers focused treatment of a range of human experiences and institutions in the world history context. The purpose is to provide serious, if brief, discussions of important topics as additions to textbook coverage and document collections. The treatments will allow students to probe particular facets of the human story in greater depth than textbook coverage allows, and to gain a fuller sense of historians' analytical methods and debates in the process. Each topic is handled over time – allowing discussions of changes and continuities. Each topic is assessed in terms of a range of different societies and religions – allowing comparisons of relevant similarities and differences. Each book in the series helps readers deal with world history in action, evaluating global contexts as they work through some of the key components of human society and human life.

The Environment in World History (Second Edition)
Stephen Mosley

Globalization in World History (Fourth Edition)
Peter N. Stearns

The Atlantic Slave Trade in World History (Second Edition)
Jeremy Black

Disasters in World History
Benjamin Reilly

Water in World History
Ellen F. Arnold

Animals in World History
Helen Louise Cowie

Animals in World History

Helen Louise Cowie

Routledge
Taylor & Francis Group

NEW YORK AND LONDON

Designed cover image: Courtesy John Lee and the National Museum of Denmark.

First published 2025
by Routledge
605 Third Avenue, New York, NY 10158

and by Routledge
4 Park Square, Milton Park, Abingdon, Oxon, OX14 4RN

Routledge is an imprint of the Taylor & Francis Group, an informa business

Library of Congress Cataloging-in-Publication Data
Names: Cowie, Helen (Helen Louise), author.
Title: Animals in world history / Helen Louise Cowie.
Description: New York, NY : Routledge Taylor & Francis Group, 2025. |
Series: Themes in world history | Includes bibliographical references and index.
Identifiers: LCCN 2024027827 (print) | LCCN 2024027828 (ebook) |
ISBN 9781032021232 (hardback) | ISBN 9781032015217 (paperback)
| ISBN 9781003181996 (ebook) | ISBN 9781040193198 (adobe pdf) |
ISBN 9781040193211 (epub)
Subjects: LCSH: Human-animal relationships--History. | Animals
and civilization. | Natural history. | Human ecology.
Classification: LCC QL85 .C69 2025 (print) | LCC QL85 (ebook) |
DDC 590--dc23/eng/20240620
LC record available at https://lccn.loc.gov/2024027827
LC ebook record available at https://lccn.loc.gov/2024027828

ISBN: 978-1-032-02123-2 (hbk)
ISBN: 978-1-032-01521-7 (pbk)
ISBN: 978-1-003-18199-6 (ebk)

DOI: 10.4324/9781003181996

Typeset in Times New Roman
by SPi Technologies India Pvt Ltd (Straive)

Contents

Figures

Acknowledgements

The research for this book was undertaken at the University of York, where I benefited from the support and friendship of my colleagues in the history department. I would particularly like to thank Dilnoza Duturaeva, who co-taught my 'Animals' module in 2023 and introduced me to some valuable work on animal history in China and Central Asia; Stephanie Howard Smith, with whom I shared many wonderful conversations about tapirs (among other things); and Hanne Cottyn, who shared her expertise on camelids and spectacled bears. I would also like to extend my thanks to several cohorts of students on my animal history module, whose seminar contributions helped to shape my work.

Beyond York, I owe thanks for a variety of individuals who have read and commented on parts of the book or provided moral support as it developed. Rachael Pasierowska offered suggestions for improving Chapters 1 and 3. Franklin Martínez and his colleagues at the Linnaeus University Centre for Concurrences in Colonial and Postcolonial Studies provided valuable feedback on Chapter 6, while Steven Navarrete Cardona kindly shared his important research on animal protection in nineteenth-century Spanish America. Participants in the Commodities of Empire workshops 'Livestock as Global and Imperial Commodities' (Berlin, 2022) and 'Fur, Fin and Feather: Commodifying Wild Animals' (York, 2023) enriched my knowledge of animals as commodities and labourers, and helped me to think more deeply about the global movement of cattle, civets and elephants – among other species. I am grateful to Samuel Coghe for organising the former event, and to the other members of the Commodities of Empire Working Group for providing feedback on an earlier chapter on the commodification of animals.

At Routledge Kimberley Smith, Emily Irvine and Allison Sambucini provided support and encouragement for the project, while series editors Peter Stearn and Jane Hooper offered helpful advice on structuring and improving it. I am also very grateful to Abel Alves, Elizabeth Hennessey and Nigel Rothfels, who read the initial proposal and provided valuable suggestions for expanding the scope of the book. Their recommendations undoubtedly enriched the project.

Last but not least, my thanks go to my parents, Peter and Susan Cowie, my sister, Alice Cowie, my parents-in-law, Gordon and Mari Williams, and, above all, my husband, Paul Williams, who very tolerantly followed me around Florence, in search of a giraffe fresco, and around Mexico City, in search of a giant panda. I would also like to pay tribute to the various feline friends who kept me company while writing this book, especially Gordon Longfellow, Jeepers Creepers, Tiger and, above all, Daisy Cowie.

Introduction

Humans have had a long and close relationship with non-human animals. They have hunted them, domesticated them, consumed them as food, exploited them as sources of energy, adopted them as companions, exhibited them for entertainment and devoted time to studying them. In the process, they have radically changed the way in which many animals live, subjecting them to captivity, altering their diets, constraining their movements and, through selective breeding, even changing the shape of their bodies. They have also impacted the overall biodiversity of the planet, driving some species to extinction and permitting others to colonise new continents.

While humans have transformed the lives of many other species, animals have played a central role in the development of human society. Oxen, horses and camels have helped humans to plough fields, win wars, engage in trade and forge empires. Llamas, jaguars and wolves have featured in religions, myths and fables. Dogs, cats, parrots and lions have served as models for painters and muses for poets, while frogs, rats and chimpanzees have functioned as experimental subjects. Though often presented as passive recipients of human actions, some animals have left their mark on the historical record, facilitating or thwarting human endeavours; in 1916, for instance, a herd of wild giraffes interfered with General Smuts's East African campaign in World War I by 'pulling down the telegraph wires and thereby interrupting communication'.[1] Less visibly, but more lethally, other animals have functioned as vectors of epizootic diseases, transmitting deadly viruses and bacteria to human populations. The Black Death, a disease spread by fleas and possibly originating in gerbils, ravaged Western Eurasia and North Africa from 1346 to 1353, killing between 30% and 50% of the continent's population and hastening the end of serfdom.[2] Malaria and yellow fever, both spread by mosquitoes, delayed European efforts to colonise parts of the Caribbean and later contributed to the success of independence movements in the United States, Haiti and northern South America by decimating thousands of unseasoned British, French and Spanish troops.[3] Animals

DOI: 10.4324/9781003181996-1

have thus shaped the contours of human history, influencing settlement patterns, diet, social structures and military encounters.

While animals have been central to human society since the earliest periods of human history, the field of animal history is much younger. Animals have, of course, made cameo appearances in a range of historical scholarship, whether as trade statistics in works of economic history, as symbolic referents in cultural histories or as colourful asides in biographies of famous historical figures. Historians of the British Civil Wars (1642–1651), for instance, often make passing mention of Prince Rupert's pet poodle, Boy, who was killed at the battle of Marston Moor in 1644.[4] Until the final decades of the twentieth century, however, the history of animals did not constitute a discrete field, and few historians granted animals a central role in their analysis. Instead, animals were viewed primarily as commodities for human consumption, specimens of natural history or metaphors for human behaviour – interesting for what they revealed about human societies, but not historical subjects in their own right.

This situation changed in the 1980s when historians began to devote more explicit attention to the complex (and changing) dynamics of the human–animal relationship. In his seminal *Man and the Natural World* (1983), Keith Thomas, showed how major social and cultural changes in the period 1500–1800 saw a conviction in man's ascendancy over the natural world give way to a new concern for the environment and sense of kinship with other species[5] Four years later, Harriet Ritvo's *The Animal Estate* (1987) shifted the focus even more squarely onto non-human animals, exploring the multiple ways in which human–animal relationships shaped Victorian Britain.[6] Since then, the field of animal history has expanded rapidly. Studies now exist on a wide range of animal-related topics, from pet keeping, to zoos, to the use of animals in science, fashion and the food industry. Historians of animals have also started to focus on the question of animal agency in history, giving greater prominence to animal actions, animal sentience and the wider animal experience. Though initially centred primarily in the anglophone world, animal history is now expanding into new areas of the globe, posing important questions about the role of animals in colonialism, urbanisation and industrialisation, and adding a non-human dimension to the histories of Spain, Brazil, the Indian Ocean and Singapore – among other places.[7] The animal turn is therefore in full swing, and animals are being written back into human history as key actors and historical subjects.[8]

Sources and Methodologies

How can we learn about human–animal relationships in the past? What sources can historians use to study animal lives in different historical contexts? What challenges do historians face in faithfully reconstructing the animal experience? What methodologies have they employed to overcome them?

Sources addressing the history of animals are by no means lacking, though they often require novel forms of interpretation. In the case of written sources, animals appear in a wide range of documents, from trade records to novels. Natural histories reveal human understandings of animals and shifting classificatory systems. Zoo guides, animal care books and inventories chart the lives of exotic animals in captivity. Hunting literature chronicles the pursuit of animals for sport, while the archives and publications of conservation organisations and animal protection societies record efforts to preserve endangered animal species and to shield individual animals from cruelty. On a more intimate level, animals frequently feature in letters, diaries, wills and inventories, and, from the early nineteenth century, in newspapers and magazines, whether as property for sale, victims of abuse, lost pets or unwelcome nocturnal visitors.[9] In 1918, for instance, the *Chicago Daily Tribune* published a story about 'a huge chimpanzee fashionably attired in a suit of men's evening clothes', who escaped from a Broadway theatre and 'strolled into the lobby of the Knickerbocker hotel'.[10]

Visual and material sources also furnish a rich vein of information about animals and can provide us with a more direct window onto the animal experience in the past. Natural history illustrations reflect attempts to depict and classify non-human species across the ages. Paintings depict animals being herded, hunted or ridden. Photography, and later film, have allowed humans to capture the movements of individual animals. Taxidermy specimens, tiger skin rugs and collections of old veterinary equipment provide insight into the lives (and often deaths) of animals in the past. Evidence of the animal past can also be found in historic buildings, in the form of cattle pens, dovecotes and cat holes, and on archaeological sites, in the form of bones, antlers, tools, figurines, rock art and – less commonly – mummified remains, providing information about how animals were butchered, what age they were when they died and what symbolic value they may have held for contemporary humans.[11] In 2019, for example, archaeologists found 100 ritually sacrificed guinea pigs at an Inca archaeological site in Peru, the majority adorned with earrings and necklaces, and some wrapped up in tiny rugs.[12] These sources – and many others – can help us to understand how animals were cared for, sacrificed and commodified in earlier societies. As Erica Fudge has pointed out, animal bodies are also sometimes present in the archive itself, in the form of vellum bindings made from calf skin.[13]

While there is no shortage of sources for studying the history of animals, accessing and interpreting them can present significant challenges. This is true, of course, of any form of history – especially histories of marginalised humans such as enslaved people, peasants, women, factory workers and societies that lacked written histories. It is, however, particularly difficult when writing the history of other species.

One major problem for historians is that sources about/containing animals can be difficult to locate. Documentation relating to non-human animals was not always perceived as worthy of retention, so a large amount of material has likely been discarded. As Susan Nance has shown, moreover, surviving documentation can often be challenging to find, as most archives were not set up with the history of animals as their primary focus. This means that horses, dogs or elephants, though they may in fact be present in the sources, do not necessarily appear as discrete entries in archive catalogues, making them hard to recover.[14]

Even when animals are present in sources, they are not necessarily all present equally. Pets, for example, tend to generate the most numerous and deepest archival traces because of their close relationship with humans. Zoo and menagerie inmates also leave behind a significant footprint, owing to their exoticism, visibility and enforced proximity to humans, while wild animals generate records when their lives intersect with those of humans in unusual (and often violent) ways; in 1898, for instance, two maneless lions in the Tsavo district of Kenya made the headlines when they killed and ate 'scores of men', forcing 'the temporary abandonment of construction work on the Uganda Railroad'.[15] By contrast, domesticated livestock tend to receive less personalised attention and are more often presented in the form of statistics, whether as entries in plantation logbooks, lists of diseased animals in public health records or rosters of horses in transport company archives. Laboratory animals are likewise 'made invisible' in the historical record, particularly routine experimental subjects such as guinea pigs.[16] What this means for historians is that some animal experiences are more accessible and recoverable than others. It will probably be easier to write the biography of a famous zoo animal like Jumbo the elephant than to reconstruct the life story of a sheep in Patagonia, a snail in Hawaii or a rat in a medical testing facility.

Of course, the big challenge when writing about animals is that they cannot write or paint for themselves, so we are always seeing them as they are refracted through human-generated lenses. This poses a major obstacle to historians, who have had to devise innovative ways of recovering the animal experience in the past. What exactly is 'animal history' and to what extent can we retrieve the voices and experiences of historical animals?

There are various possible answers to these questions. Some historians of animals focus primarily not on the animal experience per se, but on how humans have perceived, used and understood members of other species. Historians have looked, for instance, at representations of animals in the past and considered their wider cultural significance. They have chronicled shifting attitudes towards non-human animals and their welfare and they have highlighted the crucial role played by animals in many aspects of human society, from work to religion. They have also used the history of animals as a prism through which to explore broader human-centred issues, such as global trade, industrialisation and colonialism. In his essay 'The Great Cat Massacre', for instance, Robert Darnton analyses the brutal killing of cats in 1730s Paris to illustrate the mental world of the eighteenth-century French printers' apprentices who perpetrated the slaughter.[17]

Pushing this approach further, other historians have begun to explore how human societies in the past anthropomorphised and individuated members of other species, and to assess the ways in which they attempted to understand the animal experience, whether through making judgements about animal sentience and intelligence or through ascribing human-like emotions to non-human animals.[18] An alpaca shipped from Peru to Australia in 1858 was described in a contemporary newspaper as being 'very low spirited' and 'utter[ing] low plaintive moans as if in remembrance of its two companions, which died on the passage'.[19] An elephant called Alice reportedly 'trembled with fear' when she was transferred from Luna Park Circus to the Bronx Zoo in 1908, but 'squealed for joy' when she spotted her old keeper, 'carress[ing] him gently' with her trunk.[20] These projected emotions were not necessarily accurate from a biological perspective, and may have misconstrued the actual feelings of the individuals involved. They do, nonetheless, reveal what humans *thought* other animals were thinking and feeling – something which likely influenced how they treated them.

A second, more radical approach to the history of animals puts animal behaviours, feelings and actions at the centre of the historical narrative. This is a challenging exercise, for the reasons outlined above, but historians are experimenting with a range of tools and interdisciplinary methodologies to try to allow animals to speak for themselves.

First, some historians are drawing on modern studies of animal behaviour (ethology) to interpret animal actions and emotions in the historical record – sometimes correcting the interpretations imposed upon them by contemporaries. We now know that when a chimpanzee appears to be grinning, it is actually expressing fear – not happiness; that there is evidence (albeit contested) that elephants mourn their dead relatives; and that a bear described as swaying in a zoo cage is displaying stereotypic behaviour most likely induced by boredom. We also know how biological traits such as gestation periods, adaptation to particular climatic conditions and range/migratory behaviour can influence the ability of species to survive human depredation, accept domestication or endure translocation to new continents. This knowledge can help us to reinterpret animal behaviours in the past, both on an individual level and at a population level, and to better explain their interactions with humans. Eric Baratay, for instance, has attempted to reconstruct the journey of a young female giraffe from Sudan to Paris in 1827 from the perspective of the animal by applying modern knowledge about giraffe psychology and physiology to contemporary writings.[21]

As well as drawing on modern biology, historians are also paying greater attention to the testimonies of people who worked closely with animals in the past. Soldiers, for instance, enjoyed an intimate relationship with their horses and frequently came to appreciate their individual qualities. Hunters required detailed knowledge of their quarries in order to catch them, learning to track a wolf by his footprints or determine the proximity and sex of a deer by examining its excrement. Zookeepers and animal trainers experienced a sustained relationship with some of their charges, often recognising their distinctive qualities and aptitudes; Ram Brahma Sanyal, superintendent at Calcutta Zoo, emphasised the diverse food preferences of the different bears under his care in 1892, one of whom was 'extremely fond of sugarcane and juicy fruits' and another of whom 'refus[ed] to touch any food unless boiled eggs form[ed] a large proportion of it'.[22] By drawing on the surviving accounts of the humans in close human–animal partnerships, historians can gain a deeper sense of how these interactions operated in past societies and unearth further evidence of the behaviours and actions of specific animals. By tracing such relationships over an extended period, moreover, some historians are even beginning to show the ways in which humans and other species influenced one another's cultural and biological development, evolving together in response to changing social, economic and environmental pressures. Edmund Russell, for instance, argues that greyhounds and their human owners 'co-evolved' in the period 1200–1900, shaping one another culturally and (in the case of dog breeds) biologically.[23] Mahesh Rangarajan shows how humans and lions living in the Gir forest in Gujarat have 'coadapted' to one another's presence since the mid-twentieth century, changing their behaviour to co-exist in relative peace.[24]

Finally, historians have examined the question of animal agency to explore how the actions of non-humans have shaped human history. Animal agency – what exactly it is, and how far we can apply it in historical contexts – is a contentious issue, and historians disagree as to its nature and extent. Do animals need to consciously work for a particular outcome to qualify as agents, or is it enough that their actions change, facilitate or thwart human desires? Can we ever really understand the motivations of animals through reading human-produced sources?[25] Does ascribing agency to animals as individuals obscure the wider structural constrains under which they operate and disguise 'the ways in which humans have circumscribed and dominated animal life'?[26] Taking these caveats into account, however, some historians of animals are paying increasing attention to the role of non-human animals as historical actors, showcasing the ways – big and small – in which the actions of animals have shaped human lives (and their own).

First, focusing primarily on individual animals, historians have shown how some non-human actors have influenced the treatment meted out to them by humans, whether by escaping their enclosures, refusing to labour for human benefit, fleeing from human hunters or, in the most dramatic cases, attacking or killing their human oppressors. In 1887, for instance, a hippopotamus at the Jardin des Plantes attacked his keeper while he was cleaning his cage, 'sever[ing] his carotid artery with one stroke of its teeth'.[27] In 1817 a 'ravenous' caiman seized a soldier named Gamarra as he was swimming across a river in Venezuela, dragging him under the water and drowning him.[28] In 1931 a pet kangaroo called Jimmie at Corpus Christi College, Werribee, wrestled with waiter James Owen, 'str[iking] him savagely with its paws and mauling him so seriously that it was necessary for a doctor to insert several stitches'.[29] Jason Hribal interprets such behaviour as a conscious act of defiance, equivalent to the rebellions of enslaved humans against their masters.[30]

Second, taking a broader definition of animal agency, historians are beginning to explore some of the more subtle ways in which non-human actors have manipulated or obstructed human wishes by, for instance, adopting particular behaviours in anticipation of a reward, acquiescing to veterinary care that alleviates their discomfort (even if it exacerbates their pain in the short term), or engaging in small acts of rebellion that in some way ameliorate their condition. These more understated instances of subversion (or, in some cases, compliance) typically attract less attention than overt acts of violence, but nonetheless constitute examples of what appear to be conscious decisions taken by non-humans to improve their quality of life as individuals. A 'big brindle St. Bernard' dog called Romeo, for example, resisted efforts to give him medicine for haemorrhages in 1894, 'go[ing] down flat on the grass' whenever the veterinary assistant at a pet hospital in New York 'ma[de] a move to open [his] mouth' and 'cover[ing] his muzzle with his paws'.[31] By contrast, zoologist W. Lauder Lindsay reported 'many instances of war or regimental elephants in [colonial] India going regularly, day after day, of their own accord, to military hospitals to get their wounds dressed', cooperating willing with the surgeon and seemingly 'regarding the infliction of pain as a necessary or unavoidable part of the operation'; one adult male 'readily extended himself on the ground' so that the surgeon might operate and 'bore with patience even the application of burning caustic'.[32] Though minor in themselves, and falling well short of the criteria for organised resistance, such actions manifest at least a degree of awareness that certain behaviours could lead to desired outcomes.

Third, on a macro level, historians are increasingly striving to write multispecies histories that explore the (often complex) interactions between human and non-human animals across a wide variety of settings. Looking, on the one hand, at human-engineered environments, such as cities, farms or zoological gardens, historians are charting the ways in which non-human species have left their mark on the built environment, whether as food on the hoof, sources of transportation, domestic companions or vectors of disease. Chris Pearson, for instance, has shown how dogs transformed urban living in London, Paris and New York by straying, biting humans (and potentially infecting them with rabies) and defecating on pavements.[33] Focusing, on the other hand, on wider ecosystems, environmental historians are foregrounding the different ways in which animal physiologies and behaviours influence human efforts at acclimatisation, resource extraction or (more recently) conservation, dictating which species can be successfully introduced (or reintroduced) to particular environments, which are most suited to domestication and which can best survive overharvesting. Tamara Fernando shows how the pearl fisheries of Mannar, off the coast of Sri Lanka, were shaped, not only by market economics, human labour and state power, but by the actions of 'ocean currents, oysters, sharks, stingrays, jellyfish and a host of microbial and microscopic life forms'.[34] Nancy Cushing illuminates the ways in which pigs assisted colonial occupation in nineteenth-century New South Wales by straying into 'wooded areas and gullies which had become refuges for native plants and animals and for Indigenous Australians'.[35] By taking a more holistic approach to historical agency, historians are beginning to challenge the idea of human exceptionalism and to highlight the multiple ways in which animals have eluded human control – whether by evading capture in the wild, declining to breed in captivity, preying on humans and their livestock, or colonising new territories. Animals thus have agency in a historical context, though their actions range from violent rebellion and conscious disobedience through to instinctive reactions or innate biological characteristics that nuance their relationship with humans and other species.

Chronologies

While methodologies thus vary, one important thing that all historians of animals are trying to do is to historicise the human–animal relationship and to treat non-human animals as historical subjects shaped by their wider historical contexts – as opposed to static, unchanging beings divorced from the cultures in which they operated. This means thinking about how animals have adapted and evolved in direct response to human actions. It also means having an awareness of cultural differences between humans, and the ways in which these have influenced the human–animal relationship.

If we focus, for instance, on animal bodies, scholars have shown how these have changed physically over time, not just in response to gradual environmental changes (evolution), but often within a much shorter timespan as a direct or indirect result of human intervention. Dogs have developed into creatures with radically different body shapes since their domestication 15,000 years ago, while cows have been selectively bred to increase their milk and meat production. Recent biological studies have revealed that wild African elephants in Mozambique are being born increasingly frequently with smaller tusks (or no tusks at all) – a direct result of the selective pressures imposed on the species by poaching in recent decades.[36] These bodily changes, moreover, are not necessarily the result of steady, consistent intervention, happening over centuries, but can often be dated precisely to specific times and places. In the case of dogs, the precise form of most modern breeds – and, indeed, the concept of breed itself – had its origins in Britain, in the 1860s, when owners started to breed their animals for form rather than function – a process historians Neil Pemberton, Julie-Marie Strange and Michael Worboys link specifically to the contemporary rise of the dog show.[37]

As well as highlighting how animal minds and bodies have been shaped by human influences, historians of animals have also emphasised the ways in which different human societies have interacted with animals, giving cultural specificity to some of these relationships. Not every culture, for instance, has the same understanding of pet keeping (what constitutes a pet?) or meat consumption (which species should and shouldn't be eaten?), while the use of certain species has sometimes changed over time or between places; guinea pigs, for example, consumed as meat by Andean peoples, were used as pets and laboratory subjects in twentieth-century Europe and the USA.[38] Historians of animals have highlighted these shifting forms of engagement, tracing changes over time and between places. Importantly, historical studies have also proven that certain practices characterised as 'traditional' actually originated much more recently, in response to specific social and economic conditions. Jakobina Arch, for instance, has shown how the consumption of whale meat only became popular in Japan in the late 1940s, when it was put on school menus as a form of inexpensive protein in the aftermath of World War II.[39] Liz P. Chee shows, similarly, how the use of bear bile in China for medicinal purposes was not the result of a long-standing belief, but began in Jilin province in the 1980s as part of a series of economic reforms introduced by Deng Xiaoping.[40] Historicising the human–animal relationship thus encourages us to ask when major changes took place and why, unpicking the precise conjunction of factors that brought about the extinction of the Tasmanian tiger, the elevation of the panda to China's national animal or the disappearance of the turnspit dog from early modern kitchens.

Given these complexities, the task of dividing human–animal history into neat chronological periods may seem challenging, for the precise dynamics of the human–animal relationship vary significantly between places and cultures. While sensitivity needs to be shown to the specifics, however, it is possible to identify key turning points in our relationship with other species that have radically changed the way in which we perceive, treat and use our fellow creatures. Four of these stand out as particularly significant: the human domestication of other species, the accelerated translocation of animals across the globe from the sixteenth century onwards, the changes brought about by industrialisation in the nineteenth century and, most recently, the impacts of globalisation.

The domestication of non-human animals began around 10,000 years ago and had a transformative impact on the human–animal relationship. Before animals were domesticated, humans were largely itinerant, spending much of their time hunting and gathering and often migrating to follow mobile prey (such as bison, caribou or reindeer). After domestication, this situation changed. Humans in different parts of the world began, increasingly, to settle in a single place and farm, or, at least to embrace a nomadic existence based on finding pasture for their animals. Agricultural yields improved, thanks to the use of animal power for ploughing, while commerce between different human groups increased as horses, donkeys, camels and llamas were enlisted as beasts of burden. This, in turn, facilitated other social and cultural developments, from the emergence of towns and cities to the formation of organised religions – the latter often based around the worship or sacrifice of non-human animals. It also gave rise to the use of animals in warfare, as emerging states vied for power and influence across increasingly large territories. For non-human animals, meanwhile, domestication was a game changer, giving humans control over the breeding of domesticated species and intensifying the persecution of wild species that ate human crops (such as mice) or preyed on human livestock (such as wolves).

A second major turning point in the human–animal relationship came in the late-fifteenth and early-sixteenth centuries, when European overseas exploration and colonisation facilitated the mass translocation of animal species between continents. Humans had, of course, transported species across oceans before, and would do so again; Polynesian settlers brought dogs and rats to New Zealand in the fourteenth century, while British colonists introduced rabbits, sheep and camels into Australia in the eighteenth and nineteenth centuries. The so-called Columbian Exchange of the early modern period, however, constituted the most extensive interchange of fauna between previously separated landmasses, and had transformative effects on both sides of the Atlantic, bringing horses, cows, pigs, sheep and chickens to the Americas, and transporting previously unknown plants and animals to Europe, Africa and Asia, from pineapples to turkeys. This revolutionised diets on all five continents, posed classificatory challenges for early modern naturalists, and radically changed work and warfare in the Americas, where riding animals and wheeled vehicles were introduced for the first time.[41] It also triggered substantial ecological transformations across the New World, with differing impacts on different species; camelid populations, for instance, collapsed in the century following the Spanish conquest of Peru, due to over-harvesting, introduced diseases and competition with sheep; jaguars increased in number over the same period, benefiting indirectly from the drop in the human population (caused by newly-introduced pathogens) and consequent reforestation.[42] We should therefore see the early modern period as one of accelerated global interchange of biota, as elephants and rhinoceroses entered princely menageries, parrots and marmosets travelled from the New World to Europe as pets for the elite and cattle and sheep colonised vast tracts of northern Mexico.

Industrialisation marks a third transformation in the human–animal relationship, brought about, in this case, by new technologies and consumption patterns. Steam shipping, rail transportation and refrigeration made it possible to transport livestock – and later meat – over long distances, allowing Europe to outsource its meat supply to overseas territories. New methods for disassembling, processing, retailing and marketing animal products massively expanded the demand for animal commodities, from ostrich feathers to elephant ivory, while the advent of ever more deadly firearms made it easier to cull wild animals in large numbers. For working animals, meanwhile, social and technological changes triggered a surge in demand for animal power, as horses were requisitioned to transport goods and passengers around expanding cities, followed, in the early twentieth century, by a shift away from horse power and animal fuels (whale and seal oil), as petroleum became the dominant source of power for heating, lighting and motorised vehicles. Industrialisation thus reconfigured human–animal relations, changing how non-human animals were consumed and employed. The urbanisation that accompanied industrialisation, moreover, gave rise to novel forms of human–animal interaction, from an expansion in middle-class pet-keeping to the founding of the first zoological gardens. It also gave rise to the first animal protection movements, as the suffering of livestock and working animals on city streets became more intense and more visible.

The advent of intensive agriculture in the mid-twentieth century has once again recalibrated our relationship with other species, with major implications for biodiversity and animal welfare. Beginning in the USA in the 1930s (and intensifying in the post-war period), factory farming has revolutionised how we rear animals for meat, eggs and milk, moving chickens into cages, pigs into crates and cattle into feedlots. This process has been adopted on a global scale since the 1980s, with concentrated animal feed operations (CAFOs) emerging in the European Union, China and Brazil. By distancing most humans from the animals they eat, intensive agriculture has changed our relationship with other species, making supply chains increasingly complex and obscuring the link between pigs and pork, cows and beef. At the same time, however, globalisation, fuelled by new technologies such as aviation, television and the internet, has allowed humans to get closer to some animals, whether through visiting gorillas in a zoo, watching wildlife documentaries about killer whales or sharing videos of Grumpy Cat on YouTube. The encroachment of humans on animal habitats, furthermore, has accelerated the transmission of zoonotic diseases, both within and between species, posing severe risks to both human and non-human animals; the SARS epidemic of 2003, for instance, has been linked to masked palm civets, while the global Covid-19 pandemic of 2020 has been linked to bats and pangolins.[43] Since the mid-twentieth century, therefore, humans have exerted more control over animal bodies than ever before, converting billions of livestock into glorified breeding machines, decreasing the habitat available for non-domesticated species and triggering changes to our climate that will impact on all living things. More positively, the same period has witnessed advances in ethology, ecology and veterinary science, improving our understanding of animal sentience, prolonging (some) animal lives and strengthening calls for animal protection.

Themes

Each chapter of *Animals in World History* focuses on a different aspect of the human–animal relationship. Chapter 1 concentrates on the consumption of (dead) animals as food, fashion, fuel and fertiliser. Chapter 2 examines animal labour, showing how horses, mules, donkeys, camels, dogs, oxen, llamas and elephants have provided humans with power, transportation and partners in hunting and warfare. Chapter 3 explores the use of non-human animals as companions. Chapter 4 chronicles the exhibition of animals in menageries, zoos and circuses. Chapter 5 focuses on animals as sources of knowledge, both as subjects for study in their own right (as zoological specimens) and as objects of human experimentation.

The final two chapters assess how the multiple human uses of animals have impacted on their populations and on their individual well-being. Chapter 6 examines how far humans have altered global biodiversity through habitat destruction, the introduction of invasive species, pollution, human population growth, over-harvesting and (since the mid-twentieth century) human-induced climate change. Chapter 7 focuses on the issue of animal suffering and asks whether humans have become kinder or crueller to other species over time. Both chapters reflect on the complex ethics surrounding conservation and animal protection, asking why some animals have been prioritised for care over others. I also consider the relationship between animal advocacy and social justice. Has the fight for animal rights complemented, or worked against, the fight for the rights of disadvantaged humans?

Though the book is structured thematically, each chapter traces change over time and space, exploring how different types of human society interacted with animals. Hunter–gatherer societies, for example, had an intimate knowledge of their prey, gained through close co-existence, and typically treated the latter with respect. Agricultural societies lived in close proximity with domesticated animals (now viewed as increasingly as property), and relied on them for food, labour and transportation. Industrial societies became increasingly detached from the animals they consumed, while, nonetheless, consuming non-human species in unprecedented numbers as meat, pets and laboratory subjects. The book considers how shifts in human lifestyle have changed the way animals were farmed, used and perceived by humans. It also highlights significant differences between cultures, which adopted agricultural and industrial ways of life at different rates and in different ways. Japan, for example, industrialised in 1868, following the Meiji Revolution, while China only became an industrial society in the 1980s. The Aónikenks and Selk'nam of Tierra del Fuego maintained a hunter–gatherer lifestyle until the early twentieth century, subsisting on the hides and meat of wild guanacos and rheas.

Animals in World History also assesses the role of species in shaping human–animal relations and considers how traits possessed by different animals have coloured their relationship with humans. Gazelles, for instance, were unsuitable for domestication owing to their innate flightiness. Scientific experiments on baboons raised moral qualms that experiments on frogs and mice did not. Pandas, elephants and kakapos were prioritised for conservation in the nineteenth and twentieth centuries, while many insects went extinct to little fanfare. *Animals in World History* explores how the biology and behaviour of different species influenced human–animal interactions, determining whether they were domesticated, whether they were experimented on and whether they were eaten. It considers how cultural categories such as domesticated/wild, native/exotic, edible/inedible and pet/working animal have shaped human perceptions of non-human animals, with important repercussions for their treatment, value and legal status (both as species and individuals).

Notes

1 'The War in a Menagerie', *The Times*, 27 May 1916.
2 Boris V. Schmid et al., 'Climate-Driven Introduction of the Black Death and Successive Plague Reintroductions into Europe', *Proceedings of the National Academy of Science* 112:10 (2015), pp. 3020–3025.
3 John McNeill, *Mosquito Empires: Ecology and War in the Greater Caribbean, 1620–1914* (Cambridge: Cambridge University Press, 2010).
4 Mark Stoyle, *The Black Legend of Prince Rupert's Dog: Witchcraft and Propaganda during the English Civil War* (Liverpool: University of Liverpool Press, 2011).
5 Keith Thomas, *Man and the Natural World: Changing Attitudes in England, 1500–1800* (London: Penguin, 1983).

6 Harriet Ritvo, *The Animal Estate: The English and Other Creatures in the Victorian Age* (Cambridge, MA: Harvard University Press, 1987).

7 See, for instance, Abel Alves, *The Animals of Spain* (Leiden: Brill, 2011); Ana Lucia Camphora, *Animals and Society in Brazil from the Sixteenth to Nineteenth Centuries* (Winwick: White Horse Press, 2021); Martha Chaiklin, Philip Gooding and Gwyn Campbell, *Animal Trade Histories in the Indian Ocean World* (London: Palgrave Macmillan, 2020); Timothy Barnard, *Imperial Creatures: Humans and Other Animals in Colonial Singapore, 1819–1942* (Singapore: National University of Singapore Press, 2019); Sandra Swart, *The Lion's Historian: Africa's Animal Past* (Johannesburg: Jacana, 2023).

8 Joshua Specht, 'Animal History after Its Triumph: Unexpected Animals, Evolutionary Approaches, and the Animal Lens', *History Compass* 14:7 (2016), pp. 326–336.

9 Erica Fudge, *Quick Cattle and Dying Wishes: People and their Animals in Early Modern England* (Ithaca: Cornell University Press, 2018).

10 'Fainting Women Left in Wake of Chimpanzee', *Chicago Daily Tribune*, 18 February 1918.

11 Erica Hill, 'Pre-Domestication: Zooarchaeology', in Brett Mizelle, André Krebber, Mieke Roscher and Aline Steinbrecher (eds), *Handbook for Historical Animal Studies* (Berlin: De Gruyter, 2021), pp. 21–36.

12 Lidio Valdez, 'Inka Sacrificial Guinea Pigs from Tambo Viejo, Peru', *International Journal of Osteoarchaeology* 29 (2019), pp. 595–601.

13 Erica Fudge, *Perceiving Animals: Humans and Beasts in Early Modern English Culture* (Champaign: University of Illinois Press, 2002), p. 2.

14 Susan Nance, 'Introduction', in Susan Nance (ed.), *The Historical Animal* (Syracuse: Syracuse University Press, 2015), pp. 1–17.

15 'The Man-Eating Lions of Tsavo', *New York Tribune*, 21 February 1909. See also Krista Maglen, '"An Alligator Got Betty": Dangerous Animals as Historical Agents', *Environment and History* 24 (2018), pp. 187–207.

16 Joanna Dean, 'Guinea Pig Agnotology', in Jennifer Bonnell and Sean Kheraj (eds), *Traces of the Animal Past: Methodological Challenges in Animal History* (Calgary: University of Calgary Press, 2021), pp. 175–197.

17 'Workers Revolt: The Great Cat Massacre of the Rue Saint Severin', in Robert Darnton (ed.), *The Great Cat Massacre: And Other Episodes in French Cultural History* (London: Allen Lane, 1984), pp. 75–104.

18 Philip Howell describes this as 'ascribed animal agency'. See Philip Howell, 'Animals, Agency and History', in Philip Howell and Hilda Kean (eds), *The Routledge Companion to Animal–Human History* (London: Routledge, 2018), pp. 197–221.

19 'Stock by the Orient', *Adelaide Observer*, 25 September 1858.

20 'Old Keeper Brings Elephant to Terms', *New York Times*, 20 September 1908.

21 Éric Baratay, 'The Giraffe's Journey in France (1826–7): Entering Another World', in Clemmens Wischermann, Aline Steinbrecher and Philip Howell (eds), *Animal History in the Modern City: Exploring Liminality* (London: Bloomsbury, 2019), pp. 91–104.

22 Ram Brahma Sanyal, *A Handbook of the Management of Animals in Captivity in Lower Bengal* (Calcutta: Bengal Secretariat Press, 1892), pp. 103 and 99.

23 Edmund Russell, *Greyhound Nation: A Coevolutionary History of England, 1200–1900* (Cambridge: Cambridge University Press, 2018).

24 Mahesh Rangarajan, 'Animals with Rich Histories: The Case of the Lions of Gir Forest, Gujarat, India', *History and Theory* 52 (2013), pp. 109–127.

25 Chris Pearson, 'History and Animal Agencies', in Linda Kalof (ed.), *The Oxford Handbook of Animal Studies* (Oxford: Oxford University Press, 2016), pp. 240–257.

26 Specht, 'Animal History after Its Triumph, p. 332.

27 'Theatrical Gossip', *The Era*, 8 January 1887.

28 Richard Longeville Vowell, *Campaigns and Cruises, in Venezuela and New Grenada, and in the Pacific Ocean; from 1817–1830* (London: Longman and Co., 1831), p. 57.

29 'Attacked by Pet Kangaroo', *Albury Banner*, 27 November 1931.

30 Jason Hribal, *Fear of the Animal Planet: The Hidden History of Animal Resistance* (Oakland CA: AK Press, 2010).

31 'Pet Animal Hospital', *Washington Post*, 16 May 1894.

32 W. Lauder Lindsay, *Mind in the Lower Animals in Health and Disease* (London: C. Kegan and Co., 1879), vol. II, pp. 371–372.

33 Chris Pearson, *Dogopolis: How Dogs and Humans Made Modern New York, London and Paris* (Chicago: University of Chicago Press, 2021).

34 Tamara Fernando, 'Seeing Like the Sea: A Multispecies History of the Ceylon Pearl Fisheries, 1800–1925', *Past and Present* 254 (2022), pp. 127–160.

35 Nancy Cushing, '"Cunning, Intractable, Destructive Animals": Pigs as Co-Colonists in the Hunter Valley of New South Wales, 1840–1860, in Nancy Cushing and Jodi Frawley (eds), *Animals Count: How Population Size Matters in Animal–Human Relations* (New York: Routledge, 2018), pp. 113–125.

36 Shane C. Campbell-Staton et al., 'Ivory Poaching and the Rapid Evolution of Tusklessness in African Elephants', *Science* 374 (2021), pp. 483–487.

37 Neil Pemberton, Julie-Marie Strange and Michael Worboys, *The Invention of the Modern Dog* (Baltimore: Johns Hopkins University Press, 2018).

38 Edmundo Morales, *The Guinea Pig: Healing, Food and Ritual in the Andes* (Tucson: University of Arizona Press, 1995), p. 45.

39 Jakobina Arch, 'Whale Meat in Early Post-war Japan: Natural Resources and Food Culture', *Environmental History* 21 (2016), pp. 467–487.

40 Liz P.Y. Chee, *Mao's Bestiary: Medicinal Animals and Modern China* (Durham, NC: Duke University Press, 2021), pp. 139–160.

41 Alfred Crosby, *The Columbian Exchange: Biological and Cultural Consequences of 1492* (Westport: Greenwood Press, 1972).

42 Shawn William Miller, *An Environmental History of Latin America* (Cambridge: Cambridge University Press, 2007), pp. 49–76.

43 Sujit Sivasundaram, 'The Human, The Animal and the Pre-History of Covid 19', *Past and Present* 249 (2020), pp. 295–316; Matthias Glaubrecht, 'Waves of Wild Viruses', in Alex Hüntelmann, Christian Jaser, Mieke Roscher and Nadir Weber (eds), *Animals and Epidemics: Interspecies Entanglements in Historical Perspective* (Cologne: Bölhau Verlag, 2024), pp. 33–48.

Further Reading

Alves, Abel, *The Animals of Spain* (Leiden: Brill, 2011)

Arch, Jakobina, 'Whale Meat in Early Postwar Japan: Natural Resources and Food Culture', *Environmental History* 21 (2016), pp. 467–487

Baratay, Éric, 'The Giraffe's Journey in France (1826–7): Entering Another World', in Clemmens Wischermann, Aline Steinbrecher and Philip Howell (eds), *Animal History in the Modern City: Exploring Liminality* (London: Bloomsbury, 2019), pp. 91–104

Barnard, Timothy, *Imperial Creatures: Humans and Other Animals in Colonial Singapore, 1819–1942* (Singapore: National University of Singapore Press, 2019)

Bonnell, Jennifer, and Sean Kheraj, *Traces of the Animal Past: Methodological Challenges in Animal History* (Calgary: University of Calgary Press, 2021).

Camphora, Ana Lucia, *Animals and Society in Brazil from the Sixteenth to Nineteenth Centuries* (Winwick: White Horse Press, 2021)

Chaiklin, Martha, Philip Gooding and Gwyn Campbell, *Animal Trade Histories in the Indian Ocean World* (London: Palgrave Macmillan, 2020)

Crosby, Alfred, *The Columbian Exchange: Biological and Cultural Consequences of 1492* (Westport: Greenwood Press, 1972)

Cushing, Nancy, '"Cunning, Intractable, Destructive Animals": Pigs as Co-Colonists in the Hunter Valley of New South Wales, 1840–1860', in Nancy Cushing and Jodi Frawley (eds), *Animals Count: How Population Size Matters in Animal–Human Relations* (New York: Routledge, 2018), pp. 113–125.

Darnton, Robert, *The Great Cat Massacre: And Other Episodes in French Cultural History* (London: Allen Lane, 1984)

Fernando, Tamara, 'Seeing Like the Sea: A Multispecies History of the Ceylon Pearl Fisheries, 1800–1925', *Past and Present* 254 (2022), pp. 127–160

Fudge, Erica, *Perceiving Animals: Humans and Beasts in Early Modern English Culture* (Champaign: University of Illinois Press, 2002)

Howell, Philip, 'Animals, Agency and History', in Philip Howell and Hilda Kean (eds), *The Routledge Companion to Animal–Human History* (London: Routledge, 2018), pp. 197–221

Hribal, Jason, *Fear of the Animal Planet: The Hidden History of Animal Resistance* (Oakland, CA: AK Press, 2010)

Hüntelmann, Alex, Christian Jaser, Mieke Roscher and Nadir Weber (eds), *Animals and Epidemics: Interspecies Entanglements in Historical Perspective* (Cologne: Bölhau Verlag, 2024)

Maglen, Krista, '"An Alligator Got Betty": Dangerous Animals as Historical Agents', *Environment and History*, 24 (2018), pp. 187–207

McNeill, John, *Mosquito Empires: Ecology and War in the Greater Caribbean, 1620–1914* (Cambridge: Cambridge University Press, 2010)

Nance, Susan (ed.), *The Historical Animal* (Syracuse: Syracuse University Press, 2015)

Pearson, Chris, 'History and Animal Agencies', in Linda Kalof (ed.), *The Oxford Handbook of Animal Studies* (Oxford: Oxford University Press, 2016), pp. 240–257.

Pearson, Chris, *Dogopolis: How Dogs and Humans Made Modern New York, London and Paris* (Chicago: University of Chicago Press, 2021)

Pemberton, Neil, Julie-Marie Strange and Michael Worboys, *The Invention of the Modern Dog: Breed and Blood in Victorian Britain* (Baltimore: Johns Hopkins University Press, 2018)

Rangarajan, Mahesh, 'Animals with Rich Histories: The Case of the Lions of Gir Forest, Gujarat, India', *History and Theory* 52 (2013), pp. 109–127

Ritvo, Harriet, *The Animal Estate: The English and Other Creatures in the Victorian Age* (Cambridge, MA: Harvard University Press, 1987)

Russell, Edmund, *Greyhound Nation: A Coevolutionary History of England, 1200–1900* (Cambridge: Cambridge University Press, 2018)

Sivasundaram, Sujit, 'The Human, The Animal and the Pre-History of Covid 19', *Past and Present* 249 (2020), pp. 295–316

Specht, Joshua, 'Animal History after Its Triumph: Unexpected Animals, Evolutionary Approaches, and the Animal Lens', *History Compass* 14:7 (2016), pp. 326–336

Stoyle, Mark, *The Black Legend of Prince Rupert's Dog: Witchcraft and Propaganda during the English Civil War* (Liverpool: University of Liverpool Press, 2011)

Swart, Sandra, *The Lion's Historian: Africa's Animal Past* (Johannesburg: Jacana, 2023)

Thomas, Keith, *Man and the Natural World: Changing Attitudes in England, 1500–1800* (London: Penguin, 1983)

1 Consumption

Animal commodities have been central to human existence for millennia. As well as supplying human dietary needs in the form of meat, milk and eggs, they have been used for clothing, for ornamentation and for medicine. Animal fat, blubber and excrement have also served as fertiliser and as fuel, permitting humans to grow crops, cook food and heat and light their homes.

Historians have charted the exploitation of animals from a variety of perspectives. Economic historians have focused on production methods, trade statistics and supply chains, showing how animal commodities were procured, conveyed and consumed. Environmental historians have studied the impact of commercial hunting on particular species and, more recently, upon the wider ecosystems from which they are extracted. Cultural and social historians have traced the uses of animals among different cultures and societies, while historians of animals have put the animals themselves at the centre of analysis, addressing issues such as animal sentience and animal agency. In recent years there has also been a trend towards histories of individual animal commodities, tracing the global repercussions of their exploitation. Mark Kurlansky's book *Cod* (1999) charts the history of a fish whose consumption shaped communities on both sides of the Atlantic.[1] Sarah Abreyava Stein's *Plumes: Ostrich Feathers, Jews and a Lost World of Global Commerce* (2008) chronicles the rise and fall of ostrich feathers as a fashion accessory and reconstructs the networks of (mostly Jewish) merchants, labourers and financiers who facilitated the trade.[2]

Drawing on the burgeoning scholarship on animal commodities, this chapter provides an overview of the commodification of animals and explores their changing role within global trade. How did the consumption of animal commodities change over time? What role did imperial expansion, new technologies and changing tastes and fashions play in shaping the extraction and use of animal products? What ecological and ethical issues were posed by the commodification of animals?

DOI: 10.4324/9781003181996-2

Uses

Almost every part of the animal body has been used in some way by humans. Animal fur, feathers, skins and wool have provided warmth. Animal flesh has provided sustenance. Animal shells, bones and teeth have provided material for tools, trinkets and ornaments and animal fat and oil have served as fuel. Before the advent of synthetic materials in the early twentieth century, natural substances derived from plants and animals were the only ones available to humans. They were valued highly and, in some cases, traded over long distances.

From their first emergence as a species, humans have exploited animals as a source of food. Pigs, sheep, cows, chickens, turkeys, llamas, guinea pigs have been reared or hunted for their meat. Cows, goats, sheep and camels have been farmed for their milk, which has been used, in turn to make butter, cheese and yoghurt. Chickens, ducks and quails have provided humans with eggs, while bees have furnished humans with honey – the main sweetener available before the globalisation of sugar in the sixteenth century. Fish of many different species have been consumed by humans across the globe and remain a key source of protein in the diets of coastal peoples like the Inuit. Though traded mostly on a local or regional basis in the pre-modern era, meat crossed continents and oceans from the mid-nineteenth century, emerging as a key global commodity. Factory farming has also facilitated the creation of breeds tailored specifically for use as meat or dairy, enabling farmers to rear animals more quickly on less fodder.

In addition to serving as a source of protein and fat, animals have been widely consumed for medicinal reasons. Bezoar stones found in the intestines of ruminants (cashmere goats in Asia, vicuñas in South America) were believed to neutralise poison.[3] Elk hooves and tapir toenails were employed as a cure for epilepsy.[4] In colonial Spanish America the flesh from the left arm of a sloth was thought to relieve a bad heart.[5] Though some of these cures had a purely local or regional circulation, others circulated on a global basis, giving rise to extensive transcontinental trades; several thousand bezoar stones were shipped to Spain in the first decades of the seventeenth century – with devastating consequences for Peruvian camelid populations (see Chapter 6).[6] In more recent years, ostrich corneas have been used in organ transplants, while bear bile, pangolin scales and rhino horn feature prominently in 'traditional' Chinese medicine.[7] In 1980 *The Times* reported that a brewery in Taiwan was importing '2,000kg of tiger bones per year (from about 200 tigers) from Indonesia and Thailand, for the purpose of making tiger bone wine, which is retailed as medicine at $1.25 a bottle'.[8]

As well as being ingested, animals have long provided humans with clothing for their bodies and decoration for their homes. Sheep, alpacas and cashmere goats have furnished wool for textiles. Cows, ostriches and crocodiles have been used widely as sources of leather for making shoes, jackets and handbags. Silkworms have given humans silk, used for making dresses, gloves and kimonos, while jaguar skins were worn by Aztec warriors in the fifteenth century as evidence of their military prowess. Fur from a wide variety of animals has long been worn by humans across the globe, serving as both a source of warmth and a status symbol. Due to its high value and low weight, it was one of the first animal products (along with silk) to be traded over long distances, adorning elites in many cultures. The Qing emperors in seventeenth-century China wore hats fashioned from black sable pelts or black fox fur and robes trimmed with sea otter fur.[9] Fashionable women in nineteenth-century Britain and the USA wore jackets, gloves and muffs made from sealskin – shipped to London from Alaska.[10]

Other animal products have functioned primarily as fashion accessories, valued for their aesthetic qualities as much as their utility. Pearls from southern India and Venezuela were worn as necklaces by European aristocrats. Iridescent beetle wings were used to decorate jewellery in Thailand and India – a fashion later copied by Europeans. Hippopotamus tusks were fashioned into false teeth in early nineteenth-century Europe, while animal scents such as musk, civet and ambergris were worn in various periods and places as perfumes. Birds' feathers were treasured by many different cultures for their soft textures and vibrant colours, featuring in art, interior decoration and millinery. Aztec artisans created intricate artworks from the feathers of the Central American quetzal bird. French, British and North American women wore egret, ostrich and bird of paradise feathers in their hats – a grisly fashion that necessitated the deaths of thousands of birds.[11]

Beyond the ballroom and the boudoir, animal products have served as fuel and fertilisers, facilitating other human activities. Llama dung was used by the Incas to fertilise potatoes and was burned by the Spanish in the seventeenth century to power the silver smelting works at Potosí. According to British merchant Charles Ledger, without *taquia* (as the substance was called in Spanish) 'the large copper smelting establishments of Corocoro would be quite unable to exist a week … the Silver Amalgamating Works would be at a standstill, and … some two millions of human beings would be without fuel in a country where not a tree is to be met with for hundreds of miles'.[12] Another South American product, guano, collected on islands off the coast of Peru, fertilised agricultural land across the globe in the nineteenth century, enabling humans to (temporarily?) overcome the natural limits on population growth.[13] Whale, seal, walrus and even penguin oil, meanwhile, were used widely as fuel before the advent of petroleum, facilitating the use of gas lighting in Victorian cities.[14] Between 1890 and 1919, New Zealand merchant Joseph Hatch harvested around three million king and royal penguins on Macquarie Island, boiling 4,000 birds a day in an on-site refinery.[15] Animal excrement and fat have thus provided humans with food, heat and light, forming the basis of global supply chains.

Finally, on a less practical level, animals of many kinds have served humans as sources of entertainment, whether as domestic companions, zoo exhibits or circus performers (see Chapters 3 and 4). Live animals have been captured and traded across long distances to stock menageries and zoological gardens. Cockerels, bears and bulls have been baited for sport, while cats, dogs, parrots and hamsters (among many other species) have been adopted as pets. Across many different cultures animals have also been hunted as a form of elite recreation, advertising the power and wealth of those with the time and resources to engage in hunting as a leisure pursuit. The species targeted as quarries by recreational hunters have varied across space and over time, from stags in medieval England to tigers in colonial India. The symbolism and ritual of the hunt have, however, remained remarkably consistent across cultures, with the physical triumph of humans over (worthy) animal adversaries functioning as a tangible display of wealth, masculinity and monarchical/imperial potency. A painting by the court artist Tulsi from c.1590, for example, depicts the Mughal emperor Akbar assisting in the capture of cheetahs (which were taken alive to act as hunting companions), underlining the ruler's skill and bravery. A work by Diego Velázquez from c.1632–1637 shows king Philip IV of Spain hunting wild boar on horseback with dogs in a similar display of masculine prowess. Eaten in lavish banquets, converted into trophies or preserved as museum specimens, hunted animals have undergone other forms of commodification after death, serving by turns as food, fashion and art.[16]

Chronologies of Consumption

For most of their two-million-year existence humans have been hunter gatherers, living in small nomadic groups and surviving by hunting, fishing and scavenging. Animals were central to early human life, providing *Homo sapiens* with meat for sustenance, hides for clothing and bones for tool-making. Hunting was done primarily on a subsistence basis, however, and animal products were rarely stored, exchanged or traded with other people.[17]

From the Neolithic period the domestication of livestock gave humans greater control over animal populations, providing them with a more dependable supply of meat and dairy products. Sheep and goats were domesticated around 10,000 years ago in the Middle East, cattle from around 10,000 years ago, in the Middle East, India and Africa, pigs from around 8,000 years ago in the Near East and China, chickens from around 4,000 years ago in Southeast Asia and llamas from around 6,000 years ago in Peru.[18] Though a gradual and uneven process, domestication permitted humans to farm animals rather than hunt them, increasing the quantity of meat at their disposal and giving rise to new forms of human–animal inter-action, from milking cows to shearing sheep. Settled agriculture also made possible the exchange of surplus animal products – now increasingly viewed as property – facilitating trade in light, non-perishable high-value goods over long distances. Silk, for instance, was transported from China to Europe across Central Asia, cowrie shells were imported to India and Yunnan from the Maldives for use as currency, while Aztec merchants sourced jaguar skins and quetzal feathers from across Mesoamerica.[19] This regional and transcontinental trade relied on the development of two new technologies - the sail and the wheel, and was facilitated by the labour of pack animals such as donkeys, camels and llamas (see Chapter 2).

In the early modern period, the number of global animal commodities increased. Beaver pelts were procured in North America and sent to Europe to make felt hats. Pearls were sourced from the East and West Indies to adorn the bodies of European elites.[20] Vicuña wool from Peru was shipped to Spain to make shawls, stockings, blankets and gloves, while sable fur reached Western Europe from Russia. Though these commodities were traded widely, they remained luxury items, accessible only to the richest in society. Henry VIII possessed a 'gown of damask and velvet embellished with 80 sable pelts and another of black satin with 350 sable pelts'.[21]

The nineteenth century marked a major step up in intensity, as animal products became accessible for the first time to middle- and working-class buyers. Pianos with ivory keys became a staple of the bourgeois home in Europe and North America, and a marker of civilisation.[22] Feathers, furs and tortoiseshells were also traded on a global scale, making clothing, combs and accessories. Meat was traded across continents and oceans for the first time, initially as live cargo, later as chilled, tinned or frozen flesh. These products were increasingly affordable to a larger number of consumers and became necessities rather than luxuries.

The twentieth and twenty-first centuries have seen a further expansion of the trade in animal commodities, though with shifts in trade routes and pri-orities. On the one hand, the volume of animals farmed for their meat has expanded exponentially to meet growing global demand for animal protein. On the other hand, there has been a significant reduction in the demand for some other animal products as plastic, petroleum and nitrates have replaced ivory, whale oil and guano as material, fuel and fertiliser. Leather and fur continue to be worn on a large scale, while the growing demand for animal-based medicines in East Asia has created new markets for pangolin scales, rhino horn and tiger bone. Despite a move away from animal fuels, the sheer number of animals exploited is thus higher than ever before: the number of chickens killed annually rose from 6 billion in 1960 to around 50 billion in 2018.[23]

Explaining Commodification

How have historians explained the gradual globalisation of animal commodities? What factors have influenced the rise and fall of specific animal products? How has the hunting, farming and trade in meat, fur and feathers shaped human diets, fashions and commercial interactions? Several themes stand out as particularly significant.

Imperial Expansion

The exploration and (in some cases) conquest of new lands facilitated access to new and more exotic animal products, fuelling the global trade in skins, feathers and hides. From the fifteenth century, exploration and colonisation by European states opened new markets, initiating the global transfer of animals and their products. Dutch control over Papua New Guinea in the seventeenth century permitted the shipping of coveted birds of paradise to Amsterdam. Portuguese incursions into Asia gave rise to a global trade in bezoar stones, while the Spanish colonisation of Mexico and Peru brought turkeys (Figure 1.1) onto European dinner tables and guinea pigs into European homes; writing in 1621, Padre Pablo José de Arriaga remarked that the latter 'multiply so quickly that they can even be found in Rome, where I was very surprised to see them being sold publicly'.[24] The nineteenth century witnessed further expansion of imperial rule as European states colonised territory in Asia and penetrated deeper into Africa. London became the main commercial centre for the global trade in ivory, furs and feathers, while live animal dealers emerged in Europe's major port cities (see Chapter 4). Europeans travelling abroad – especially soldiers and sailors – also brought exotic animals home from distant lands, creating an informal market for exotic pets; one ex-Boer War combatant advertised 'Jennie, the nicest baby baboon ever brought from Cape Colony, very tame, affectionate, most amusing, clever, healthy and hardy, write for photograph and fullest particulars'.[25] The sourcing of animal products was thus closely connected to patterns of inter and intra-imperial trade, facilitated by economic and political control over distant territories.

Perhaps even more important than the access it provided to wild animals, imperial expansion released new lands for the farming of domesticated species, precipitating a massive increase in global livestock populations and allowing Europe to, in effect, outsource much of its meat, leather and wool production to other continents. The Spanish and Portuguese colonisation of America in the fifteenth and sixteenth centuries saw the introduction of cows, pigs, sheep, horses and chickens to the continent. Surrounded by an abundance of food and relatively free from predators, many of these species multiplied rapidly, ranging across the immense grasslands of

Figure 1.1 Turkeys on sale in a market in northern Italy. Detail from Jacopo Bassano, Great Market, *c*.1580.

Source: Musei Reali Torino/Superstock.

Argentina, Venezuela and northern Mexico and providing a source of meat, leather and tallow for local consumption and export to Europe. In 1587, New Spain (modern-day Mexico) exported 64,350 hides to Spain, while the Viceroyalty of the Rio de la Plata (modern-day Argentina, Uruguay, Paraguay and Bolivia) was exporting 1 million hides annually by the end of the eighteenth century.[26] In the nineteenth century, the colonisation of additional territories in Australia, New Zealand, Patagonia and the Great Plains of the USA opened up further regions for cattle ranching and sheep farming, giving rise to a global trade in beef jerky, wool, and, once the technology permitted, fresh beef and lamb. The expansion of livestock populations thus increased meat consumption globally, altered ecosystems locally and introduced new farming practices to indigenous Americans, many of whom encountered domesticated livestock for the first time. A watercolour from eighteenth-century Peru shows Indians from the northern province of Trujillo milking cows – an operation never performed on native llamas and alpacas (Figure 1.2).

Figure 1.2 'Indians milking cows', from *Trujillo del Perú*, Vol. II, Plate 79. Patrimonio Nacional, Palacio Real-Biblioteca, Madrid, Spain.

Source: Palacio Real, Madrid.

While imperial expansion acted as a catalyst for the circulation of animal commodities, it should be noted that the relationship worked in both directions, as growing demand for certain animal products encouraged hunters and traders to push forward commodity frontiers. The desire for beaver pelts, for example, incentivised French and British incursions into Canada and reshaped relations between indigenous peoples like the Hurons and Iroquois.[27] The quest for green turtles (valued for their meat) and hawksbill turtles (valued for their shells), motivated Caymanian turtle hunters to extend their range across the Caribbean Sea in the eighteenth century, travelling as far as Cuba, Jamaica and the Central American coasts.[28] The burgeoning European demand for ivory in the nineteenth century encouraged Omani traders like the notorious Tippu Tip to expand their influence further into the African interior.[29] Animal products therefore shaped trading routes, influenced the fortunes of settlements and transformed relations between indigenous peoples, hunters, traders and settlers. The produce of some animals, moreover, facilitated the cultivation of other global commodities and the exploitation of human populations elsewhere. Beef jerky from Argentina and salted cod from New England fed enslaved labourers on Caribbean and Brazilian sugarcane plantations in the eighteenth and nineteenth centuries, contributing to the global commerce in sugar.[30]

The expansion of pasture for domesticated animals served as an equally strong catalyst for colonisation, encouraging the seizure of indigenous lands for livestock grazing. In nineteenth-century North America, the desire to appropriate the prairies for cattle ranching prompted a protracted and genocidal campaign against the Plains Indians and the native bison to open the region for beef production. The Sioux and the Cheyenne were forced onto reservations in the 1870s and 80s, while the bison were exterminated, partly to deprive the Indians of their primary food source (see Chapter 6).[31] In Patagonia, during the same era, General Julio A. Roca carried out a brutal military campaign against the Mapuche, Aónikenk and Pampas Indians of Patagonia – the euphemistically named 'Conquest of the Desert' (1878–1885) – to open southern Argentina for sheep farming. Some indigenous people ended up working as peasants on cattle and sheep ranches in their former lands; others were sent north to the sparsely populated Chaco, where many perished from disease. Overall, an estimated 2,500 indigenous warriors were killed during the campaign itself, while a further 10,000 were taken into captivity to work as agricultural labourers or domestic servants.[32] This displacement process continued into Tierra del Fuego in the 1890s, with the removal/killing of the Selk'nam people to make way for Corriedale sheep, imported from New Zealand.[33] Livestock expansion was thus both a beneficiary from, and a direct cause of, external and internal colonisation, and a major driver for the displacement of native animals and peoples. As Virginia DeJohn Anderson has shown, furthermore, animals themselves could act as unwitting agents of conquest, straying beyond European settlements in seventeenth-century New England and Chesapeake Bay and damaging the Indians' crops.[34] Today, the ever-increasing global demand for meat continues to be the cause of encroachment on indigenous lands, especially in Central and South America, as new territory is sought for pasturing cattle and, increasingly, for growing soya and corn to sustain cattle in feedlots in the USA and China.

Technological Innovation

New technologies have likewise facilitated the extraction, processing and transportation of animal commodities, making them cheaper and more accessible. Faster and more regular shipping (steam-powered from the mid-nineteenth century), sped up the circulation of animal products, opening new territories for exploitation. Developments in storage and preservation allowed perishable products to last for longer and to reach new markets, while new ways of processing fur and fibre increased the volume of animal commodities manufactured. South African farmer Arthur Douglas, for instance, invented an incubator for ostrich eggs in the 1860s, bolstering the trade in ostrich feathers. Bradford wool magnate Titus Salt patented new machines for spinning alpaca wool in the 1830s, increasing demand for the product and giving rise to a network of British wool traders in Arequipa. New techniques for dehairing and dyeing sealskins improved their quality and enhanced their popularity from the 1870s.[35] Industrialisation thus accelerated the demand for many animal commodities and made them available to a mass market for the first time. More recently, by contrast, new technologies have served to decrease the use of some animal substances by creating synthetic substitutes. Plastics have replaced ivory, petroleum has replaced whale, seal and penguin oil and nylon and polyester have provided a synthetic alternative for animal-based fabrics such as fur, silk and wool.

The influence of technological innovation on animal products is perhaps most visible in the meat industry, which became fully mechanised (and consequently globalised) by the end of the nineteenth century. The expansion of the railway network, the advent of mechanised slaughter and the development of new canning and refrigeration technologies facilitated the transcontinental and intercontinental trade in dried, tinned, chilled and frozen meat, allowing consumers in Europe and the East Coast of the USA to dine on beef from the Great Plains or Argentina and lamb from New Zealand or Australia. This in turn gave rise to new cities and towns which operated as hubs for meat production. Cincinnati emerged as a major pork production centre in the mid-nineteenth century, while Chicago rose to prominence in the 1880s on the back of its meat-packing industry.[36] Buenos Aires in Argentina and the small town of Fray Bentos in Uruguay also developed into major meat processing centres, the former as an exporter of chilled and frozen beef, the latter, from the 1860s, as a pioneer of Liebig's meat extract and tinned corned beef. New technologies thus permitted meat production to be outsourced from Europe to overseas colonies, increasing the speed with which cattle were converted into steaks and significantly raising the amount of meat consumed by all social classes. The growing geographical separation of farm animal from consumer, coupled with the parallel movement of slaughterhouses to the edge of cities (enacted primarily as a public health measure) served, at the same time, to distance consumers from the animals they ate, severing the direct connection between pigs and pork, cows and beef, sheep and lamb.[37] This phenomenon – referred to by Richard Bulliet as post-domesticity – changed the relationship between consumers and their food, obscuring the animal origins of many meat products and making animal suffering easier to ignore; a processed chicken nugget wrapped in cellophane, was (and is) difficult to connect to a living chicken.[38]

While new machinery facilitated the processing and global distribution of animal products, new breeding technologies reshaped animal bodies, aligning them more closely to human needs. Beginning between 14,000 and 27,000 years ago (in the case of dogs) and intensifying from around 10,000 years ago (with sheep, goats and cattle), domestication revolutionised the human–animal relationship, converting animals from prey into property. Scholars disagree over how and why domestication came about. Some argue that it was an intentional process, prompted by the desire to improve yields of meat, milk or wool. Others suggest that it came about accidentally, as a byproduct of keeping certain species for sacrifice or other religious purposes.[39] Recent studies have also emphasised the role that some animal species may have played in their own domestication, presenting the process as a form of co-evolution; wolves, for instance, likely gathered around human encampments to scavenge on prey, while wild cats converged on human grain stores to hunt rodents.[40] Whether intentional or accidental, the results of domestication were dramatic for both humans and animals, changing human diets, settlement patterns and social structures and effecting a range of morphological changes to animals. Pigs, for example, developed shorter snouts, curly tails and (in some cases) a white colouring following domestication, while sheep lost the ability to seasonally shed their wool. Dogs developed floppy ears, shorter legs, smaller brains and a wider range of coat colours than their lupine ancestors, evolving into breeds as distinct as the saluki (an Arabian sighthound), the Pekingese (a Chinese toy dog) and the Akita Inu (a Japanese hunting dog). *Xoloitzcuintles*, bred in pre-Columbian Mexico for food and sacrifice, lack hair and have fewer teeth than other dogs, owing to a genetic mutation.[41]

In the last two centuries this process of reshaping animal bodies has accelerated as humans have applied more sophisticated biological knowledge to livestock reproduction. From the late eighteenth century some farmers, motivated by the ever-growing demand for meat, sought to 'improve' their livestock through artificial selection, often inbreeding cows and sheep to perpetuate desirable traits. The Leicestershire farmer Robert Bakewell, for instance, produced a series of rather grotesque livestock breeds, including the Leicester sheep and the Dishley longhorn, all notable for their high flesh-to-bone ratio.[42] As the nineteenth century progressed, domestic livestock were subjected to increasingly intensive selection, giving rise to fewer, more specialised breeds. Global livestock also became ever more standardised, as prized British cattle and sheep were exported across the world; shorthorn and Hereford bulls were imported to Argentina and the USA in the mid-nineteenth century to 'improve' the longhorn cattle originally imported by the Spanish.[43] The twentieth and twenty-first centuries have seen a continuation of the intensification process with the use of fortified foodstuffs, hormone injections, artificial insemination and antibiotics ushering in an era of factory farming. Animals are bred to mature more quickly, producing the most meat or milk possible in the shortest amount of time at the lowest possible cost - with major environmental and welfare implications. The poster-child for factory farming, the broiler chicken, puts on weight so quickly that it can be slaughtered after are just 42 days.[44] By this time it is so fat that its legs cannot support it.

Taste

Consumption of animal products has also been dictated by cultural preferences and religious taboos – particularly the consumption of meat. Judaism and Islam forbid the ingestion of pork, viewed as dirty, while Hindus do not eat beef, regarding cows as sacred. Christians in early medieval Europe abjured the consumption of carrion, which was seen as unclean.[45] Nor is species/origin the sole determinant of edibility, for the way in which meat is prepared can also serve as a marker of cultural identity. The Peruvian Jesuit Antonio León Pinelo, writing in 1650, criticised the Charrua people of the Rio de la Plata (modern-day Argentina) for subsisting on 'raw meat and 'drinking the blood' of recently slaughtered cows and horses.[46] An article in the *Bristol Mercury* from 1835 branded a man named Daniel Rawlings a 'voracious beast' after he consumed 'in the space of 40 minutes, a raw pig' and, later, 'a raw hedgehog, with its skin and bristles, for a trifling wager'.[47] Food taboos have therefore shaped cultural interactions and become markers of ethnic, religious or national identity. On occasion they have even triggered rebellions; when the British army issued bullets greased with cow and pig fat to Hindu and Muslim soldiers in India the latter mutinied, sparking the Sepoy Rebellion of 1857.

The case of horsemeat exemplifies these cultural predilections. Largely rejected in early modern Europe, horsemeat started to be promoted by French scientists Isidore Geoffroy de Saint-Hilaire and Emile Decroix in the 1860s as both a valuable supplement to the diet of French men and women, and (perhaps more surprisingly) a humane measure for horses, who would thus be spared a lingering death as working animals.[48] The campaign proved comparatively successful and horsemeat began to be consumed in France with some regularity; by the early twentieth century, it comprised 11.2% of French weekly meat consumption.[49] In nineteenth-century Britain, by contrast, a similar push to add horsemeat to British diets generated visceral opposition from the contemporary press, which responded to the move with disgust (and crude racial stereotypes). An article in the *Belfast News-Letter* meditated that while 'We may admire, at a distance ... the frugality of the Chinese, who do not object to fricasseed rats, curried cats and jugged up dogs ... here we are not compelled to such mean resources ... nor seduced by a curious refinement of taste to slaughter the most noble, beautiful and gifted of all inferior animals for the table'.[50] Cherished for serving humans as a mount on the battlefield, horses were viewed as too noble to be slaughtered for their meat and unsuitable for human consumption. Britain's access to a ready supply of other meats from its colonies, moreover, was seen as a marker of national esteem, rendering experiments with horsemeat unnecessary. These prejudices have endured into the twenty-first century, as evidenced by the British horsemeat scandal of 2013.[51]

Clothing and other forms of adornment have been less strongly influenced by cultural factors than meat consumption, which lies at the very core of human identity (as the saying goes: 'you are what you eat'). Fashion choices have, nonetheless, evolved over time, with certain animal commodities rising and falling in popularity in response to shifting preferences and social conventions. Animal perfumes such as civet and musk, for instance, were popular in early modern Europe, but lost their appeal in the mid-nineteenth century as consumers opted for milder floral fragrances.[52] Tortoiseshell acquired new popularity in seventeenth-century Japan due, in part, to a change in women's hairstyles, which required tortoiseshell combs to hold them in place.[53] Sumptuary laws, meanwhile, have been imposed in many different societies to enforce class hierarchies, limiting the wearing of high-status luxury products to a privileged elite. In seventeenth-century China wearing fur acted a status symbol, with the highest-ranking elites entitled to don 'Manchurian pearls, sable and lynx' and 'lower noblemen dressed in squirrel and weasel'.[54] In Inca Peru the nature of one's dress likewise denoted one's place in society, with the finest fabrics reserved for the emperor and his family. According to Jesuit missionary Bernabe Cobo, who visited Peru in the seventeenth century, clothing made from llama wool was worn by the lower classes, while fabrics made from alpaca wool were worn by the nobility. The clothes of the Inca 'were made entirely or partially of vicuña wool', mixed with 'viscacha wool ... and bat fur ... which is the most delicate of all'.[55]

From the 1870s a new factor, seasonal fashion, also came into play. With styles now changing on an annual or even monthly basis, customers increasingly desired novelty in their purchases, often prioritising this over utility and quality. In 1925, for instance, *The Times* reported that colobus monkeys from East Africa were being 'relentlessly pursued for their skins', which 'have attracted the notice of fashion' – even though they offered 'little protection against cold'.[56] In 1912, meanwhile, one commentator on the sealskin industry remarked, tellingly, that 'people do not pay for what is best; they pay for what fashion demands. If the fashion should demand baby seals, they would have to be taken'.[57] Globalised supply chains flexed quickly to meet these sudden increases in demand, intensifying farming processes or stepping up hunting. In 1892 a revival in the fashion for ermine was quickly communicated to Chinese trappers, who 'set to work anew' and rapidly supplied the London market with 'ten thousand skins'.[58] This, of course, put intense pressure on stoat populations, which – like many other species – found themselves at the mercy of the whims of consumers in distant lands. Fashion has thus become a key determinant of the demand for animal products, assisted, increasingly, by the rise of advertising, fashion magazines, department stores and, in the twenty-first century, internet influencers.

Ethical Considerations

Lastly, in assessing the factors that have curbed or accelerated the consumption of animals, it is important to factor in ethical considerations. In the case of meat, consumers have in many instances elected not to eat certain animal products on moral grounds. British expatriates and elite Chinese humanitarians in mid-twentieth-century Hong Kong, for instance, campaigned against the consumption of dog meat, for both ethical and public health reasons, arguing that it was cruel and that dog-eating was connected to rabies.[59] Since the late nineteenth century, meanwhile, animal welfare organisations have condemned the live transport of cattle, cruel forms of slaughter and the eating of foie gras (produced by force-feeding geese to enlarge their livers), claiming that such practices entail excessive suffering. As early as 1910 animal lover Lettice McNaughten was urging fellow readers of the humanitarian periodical *The Animals' Friend*, to think carefully about their dietary choices, and to shop as ethically as possible to minimise animal pain:

> See that your meat, if it is impossible to keep house without using any, is procured from a butcher who kills the animals humanely – i.e. stuns them before the knife is used ... Encourage in any way you can the establishment of public slaughter-houses. In these places the very best methods of killing can be enforced and the general treatment of the animals supervised by reliable men ... Avoid poultry and game as food, as great cruelty takes place in supplying them for the market. Geese, fowls, etc. are often tightly packed without food and water. The artificial cramming practised in many poultry farms is an exceedingly cruel and disgusting proceeding.[60]

McNaughten concluded her appeal by imploring her readers to 'give up eating flesh food altogether' if they possible could.[61] While the complete renunciation of meat has always been the preserve of a minority, vegetarianism (and, more recently, veganism), has become increasingly common in the West since the eighteenth century, with organised vegetarian movements emerging in Britain and the USA from the 1840s and the first vegetarian restaurants and cook books coming into existence.[62] Though not all vegetarians abstain from meat for ethical reasons – some simply cannot afford it – many have chosen to do so out of religious conviction (in the case of the

Jains of India), because they have concerns about the welfare of farmed animals, or, increasingly, because they are worried about the environmental impact of industrialised meat production, which is responsible for pollution, rainforest destruction and the emission of vast quantities of greenhouse gases. Some more militant vegetarians have also viewed the consumption of animal flesh as part of a wider system of capitalist exploitation and corruption, involving the abuse of farmers and slaughterhouse workers as well as animals themselves. Early twentieth-century humanitarian Henry Salt, for instance, described meat-eaters as 'almost literally cannibals, as devouring the flesh and blood of the non-human animals so closely akin to us, and indirectly cannibals, as living by the sweat and toil of the classes who do the hard work of the world'.[63] Following the advent of factory farming in the twentieth century the trend towards vegetarianism has accelerated, with growing numbers of consumers recognising the suffering and ecological damage being done by commercialised agriculture. The rejection of meat is especially strong in the anglophone world, where fast-food culture is at its most intense (though the highest number of vegetarians is to be found in India, mostly for religious reasons).

As for the consumption of non-edible animal goods, this has also been shaped, to a degree, by ethical considerations, whether related to human morality, ecological damage or animal suffering. On the one hand, the over-harvesting of wild animals for clothing or adornment has generated concerns about animal population decline and even extinction. Writing in 1895, for example, Margaretta Lemon, a founding member of the Royal Society for the Protection of Birds, condemned the slaughter of rare birds of paradise for their feathers, predicting that 'this ruthless destruction, which merely panders to the caprice of a passing fashion, will soon place one of the most beautiful denizens of our earth in the same category as the Great Auk or the Dodo'.[64] On the other hand, the violence inherent in shooting, trapping, poisoning or farming animals for their fur, horns or shells has triggered anxieties about cruelty, which certain forms of animal slaughter undoubtedly entail. In 1897, for instance, British humanitarian D. Harrison accused the Hudson Bay Company of poisoning bears with strychnine to make 'the fur glossy' leaving the animals 'doubled up' in 'agony'.[65] A desire to combat the parallel problems of species decline and animal suffering has prompted some consumers to reject fur and feathers in their attire and to adopt plant-based or synthetic alternatives in their place. Fears over extinction have also encouraged governments to enact localised prohibitions and international trade bans, limiting the legal trade in vulnerable species (see Chapter 6). Ethical considerations have therefore shaped the consumption of animal skins, wool and bones at the individual, national and international levels, especially over the last two centuries.

Although moral concerns have played a part in influencing consumer choices, it should be noted that they have probably had less overall impact than more basic issues such as quality, accessibility and price. Consumers want cheap, fashionable products, and are often willing to overlook the destruction and cruelty behind them, especially when this takes place far away or behind closed doors. As Joshua Specht has shown in relation to industrial beef production in late-nineteenth-century North America, most consumers were more bothered about the purity and price of the meat they received than the suffering of the animals or the poor working conditions endured by (often immigrant) labourers employed in Chicago's slaughter-houses, protesting only when adulteration fears were rife or when the cost of beef rose too high.[66] Similar concerns about the quality, quantity and purity of cows' milk influenced breeding programmes, production processes and public health initiatives in colonial Burma, where Indian cattle and dairy-men were viewed as a potential vector of disease.[67] While some individuals and groups have altered their consumption practices for ethical reasons, moreover, their efforts have, to an extent, been cancelled out by the emergence of alternative markets elsewhere. Increasing numbers of westerners are becoming vegetarian or vegan, but growing economic prosperity in countries such as China and India has led to a huge rise in meat consumption, which is regarded (as it has long been in the West) as a marker of social status; in China, for instance, meat consumption rose from 4 to 61 kilogrammes per capita between 1961 and 2010.[68] The exponential growth in the global population from around one billion in 1800 to 7.7 billion in 2019 has further exacerbated this trend, raising the consumption of animal commodities to ever higher levels.[69] While humanitarian and environmental campaigners can make (and have made) a difference to consumer choices, therefore, the overall trend in the consumption of animal products has been ever upwards, especially the consumption of domesticated animals for food.

Global Animal Commodities

How have the factors outlined above impacted the global consumption of specific animal commodities? To explore the shifting patterns of animal consumption in more detail, the chapter concludes with case studies of four different animal products: pork, ivory, musk and cochineal. It assesses how their use has changed over time and charts the commodity chains and global exchanges that facilitated their consumption on a global scale. It also discusses the ethical and environmental issues raised by the production of animal commodities.

Pork

Pork is derived from a domesticated species – the pig. Along with its by-products, sausages, bacon and ham, it is widely consumed across multiple cultures, appearing in Spanish chorizos, Italian salami, Mexican carnitas and British pork pies. Living pigs have also become global commodities, acquiring their current form through a series of exchanges between Asia, Europe and the Americas.

Pigs were domesticated around 7,000BC in both southern China and the Near East, in what zoologists believe to have been two separate domestication events. As omnivorous animals, they served primarily as storage vehicles for waste food products, permitting societies across Europe and Asia to feed surplus crops to their hogs and consume them later as meat. While pigs fulfilled similar functions in both Europe and Asia, they evolved differently in the two domestication sites due to contrasting social conditions. In China, which was more densely populated than Europe, pigs were generally kept in sties and allowed to forage on scraps. Relieved of the need to fend for themselves in the wild, their form and physiology gradually changed, giving rise to smaller, pot-bellied animals that matured more quickly and bred at an earlier age. In Europe, by contrast, pigs were accorded greater freedom and generally allowed to roam the forests consuming acorns and beechnuts (a system known as pannage) before being rounded up and slaughtered. Obliged to compete for food with wild boars (with which they sometimes interbred), they were more subject to natural evolutionary selection pressures than their Chinese counterparts, and, in consequence, remained agile and lean with long legs, bristled backs and residual tusks. Where Chinese pigs provided a crucial source of meat for urban communities, therefore, European pigs functioned mainly as a source for peasant subsistence, and were farmed in a less intensive way.[70]

At various points in European history humans attempted to selectively breed pigs to increase their meat output, but early experiments failed to achieve lasting change. The Romans, for instance, appear to have taken some steps towards porcine improvement, but their work was undone by the disintegration of the Roman Empire in 476AD. Later efforts at improvement in Britain and the Low Countries in the late Middle Ages were likewise thwarted by the Black Death (1346–1353), which reversed earlier human population growth and removed the need to maximise agricultural production. In the eighteenth century, however, some farmers in Britain made a more concerted effort to (as they saw it) 'improve' the domestic pig, paying greater attention to diet and breeding. Taking advantage of the Agricultural Revolution, which increased the range of fodder available for pigs, and of increased urbanisation, which gave pigs access to the waste products of breweries and dairies, breeders managed to create larger breeds which converted food into flesh more quickly. Crucially, the importation of Chinese pigs into Britain sometime around 1700 also permitted agricultural 'improvers' to cross-breed these more rotund animals with the improved British stock, giving rise to faster-breeding, quicker maturing and much fleshier pigs with a much higher fat to bone ratio.[71] Over the course of the nineteenth century these breeds were further perfected, reaching grotesque proportions; two nine-month-old 'middle white pigs' exhibited by George Hayter at the Smithfield Cattle Show in 1901 sport bulging torsos and rotund bellies, perched on short legs that seem barely capable of supporting their weight (Figure 1.3).[72] On the other side of the North Sea, meanwhile, breeders in Denmark responded to a growing desire for leaner, flecked (or marbled) meat by crossing native Jutish pigs with English Large Whites to produce the Danish Landrace, initiating the famed Danish bacon industry.[73]

The next phase of the pig's development took place in the nineteenth-century USA, where a further transfer of domesticated livestock combined with ample land and new technologies to industrialise pork production. Originally imported to America in the fifteenth and sixteenth centuries by the Spanish, pigs (the European variety) thrived in the New World, where they were consumed as meat and used as a source of lard (pork fat) for cooking. With land available in abundance, many went feral. In the nineteenth century an injection of Asian breeds (probably via Europe), coupled with increasing urbanisation on the East Coast, encouraged more intensive forms of pork production, especially in the Ohio Valley region, where farmers fed surplus corn to pigs to service expanding urban food markets (it was easier to transport pigs than corn over long distances). Cincinnati emerged as a

Figure 1.3 'A Pretty Couple', *The Graphic*, 21 December 1901.

Source: photograph by author.

national centre for pig slaughter and meatpacking, earning the nickname 'Porkopolis'. It was subsequently superseded by Chicago, the dominant meat-packing city in the nineteenth-century USA. As with beef, nationwide distribution of pork was facilitated by railroad expansion and the introduction of refrigerated rail cars, which allowed pork products to be transported over long distances. Cincinnati's pork packinghouses, meanwhile, pioneered the concept of the modern production line, using unskilled labour to disassemble pig carcasses in the most efficient manner possible.[74]

The twentieth century has witnessed the intensification of these processes, speeding up pork production still further and expanding its consumption. New technologies such as trimmers, rollers and pressers have facilitated the production of highly processed meats like peperoni and hot dogs, expanding the consumer base for pork-based products. Enhanced genetic knowledge has permitted farmers to select pigs explicitly for leanness, producing larger quantities of the most lucrative meat, while giving piglets antibiotic and vitamin-fortified food has allowed farmers to wean them earlier, enabling sows to be impregnated just nine weeks after giving birth. Partly because the new, leaner breeds of pig are less hardy than their forest-roaming ancestors, pig production has increasingly moved from farmyards into large indoor facilities, where pigs are kept in stalls and fed on large quantities of soybeans from South America. This, of course, has severe welfare implications for the pigs, which are denied the kinds of social interactions they experienced in less intensive settings.[75] It has also fuelled deforestation in Paraguay and the Brazilian Cerrado. With pork production now largely hidden from view, however, the fate of most pigs is concealed from the wider public, who are increasingly distanced from the animals that provide them with meat and struggle to associate the processed sausage with the living pig.

The story of pork production is thus one marked by global exchanges, selective breeding and technological innovation, which, together, have transformed porcine bodies and human diets across several continents. Bringing the globalisation process full circle, intensive farming techniques and fast-growing Western pig breeds have now been introduced into China, contributing to a massive increase in pork consumption in that country since 1978. In 2016 China produced 53 million tons of pork from a domestic herd of 671 million pigs; double the amount produced in the European Union and five times that produced in the USA.[76]

Ivory

Ivory was sourced from a wild animal rather than a domesticated one. In common with pork, however, it was a highly valued and widely used substance, spawning complex supply chains across Asia and Africa. Like pork, ivory also gave rise to serious ethical problems relating, in this case, to the treatment of humans involved in the trade and to the extermination of elephants.

Ivory has an interesting history. For many centuries, the material had been used by people in Africa and Asia as a form of ornament. It was carved into art and religious objects and used mostly for decorative purposes. The Roman Emperor Caligula is said to have given his horse, Incitatus, an ivory manger!

In the nineteenth century, ivory shifted from being a medium for high art in the East to a substance for everyday use in Europe and North America. Valued for its smoothness, hardness and durability, ivory was used to make cutlery handles, buttons and mathematical instruments, as well as being ground down to a powder and used as a stiffener for jellies. Most importantly, ivory was the material of choice for making billiard balls and piano keys – two items in high demand in Europe and North America and, in the case of pianos, a status symbol for the bourgeois home. It was, in effect, the plastic of its day.

Where did ivory come from? Some ivory – though only a small amount – came from Asia. Here, elephants were captured and used as workers on timber plantations. The tips of their tusks were sawn off periodically (about every 8–10 years, according to one account), and sent to Britain for processing. Most ivory, however, emanated from Africa, where it was procured through contacts with local people. Tusks brought from Cameroon were transported across the desert on the backs of camels by Hausa traders. Tusks brought from East Africa were sourced from indigenous elephant hunters (known as *Makua*) and transported to trading hubs such as Zanzibar in caravans. The latter were organised by Swahili and Omani merchants like the slave trader Tippu Tip, who became wealthy from the profits of the ivory trade. By 1891 the island of Zanzibar off the coast of German East Africa (modern-day Tanzania) was supplying 75% of the world's ivory, much of it re-exported to Europe from Mumbai. Most of this ivory ended up in London, where it was sold at auction to piano makers and cutlery manufacturers.[77]

Ivory was thus a global commodity with a large consumer base. Its consumption, however, posed serious ethical issues, giving rise to heated debates towards the end of the nineteenth century. These debates concerned both the human costs associated with sourcing ivory and the environmental impact of the ivory trade.

One major issue with the African ivory trade was that it was intimately connected with the slave trade. In the mid-nineteenth century, a lot of the ivory imported into Europe came not from freshly killed elephants, but from large stockpiles that had been built up by sub-Saharan African communities over the centuries. As the value of ivory increased, Arab middlemen saw that they could make a quick profit by raiding these villages and requisitioning their ivory. In the process, they often enslaved the villagers, who were impressed to carry the stockpiled ivory to the coast for export. This meant that the demand for ivory in Europe and North America directly promoted slavery, undermining the aims of Christian missionaries. As General Charles Gordon put it, the ivory trade was 'only the slave trade under another name'.[78]

Second, the ivory trade had devastating impacts on the elephants themselves, undermining its long-term sustainability. Once all the readily available stockpiles of ivory had been exhausted, hunters started to slaughter living elephants for their tusks. This was not sustainable as elephants breed very slowly and the volume of tusks exported was high; in 1882, one Sheffield cutler was getting through 522 tusks in a fortnight (for knife handles) – necessitating the slaughter of at least 261 elephants.[79] The persistent slaughter of elephants meant that some regions of Africa became completely depopulated of the species, giving rise to fears that they would soon become extinct. A journalist writing in *The Graphic* warned that 'if matters go on much as they are now, this noble quadruped, the majestic living reminder of the old days, when frogs were as big as bullocks and elks towered like giraffes, will become as extinct as his hairy-coated brother, the mammoth'.[80]

Anxious to save the elephant from this sorry fate, contemporaries enacted a range of measures, all aimed at reducing the supply of, and demand for, ivory. Addressing the supply side of the equation, some colonial administrators introduced hunting licenses in their territories or established game reserves where elephants could not be hunted (at least in theory). Several European powers also implemented bans on the export of underweight ivory, discouraging the hunting of female and adolescent elephants. Addressing the demand element of the problem, scientists tried to find alternatives to ivory that might replace it as a manufacturing material. Some suggested using other forms of animal ivory, from walrus tusks to hippopotamus teeth. Others recommended a substance known as vegetable ivory, which came from a type of nut grown in South America. Yet others experimented with synthetic substitutes for ivory, such as celluloid. Ultimately, the invention of plastic in the early twentieth century came to the elephant's rescue, replacing ivory in most of its key functions.[81]

Since the mid-twentieth century, demand for ivory has risen again, with poaching resuming in the 1970s and 80s to meet a growing market in the Far East – initially in Japan and Hong Kong, and now, increasingly, in China. As a result, African elephant populations have plummeted from around 2 million in 1973 to just 415,000 in 2017. Conservationists have responded to the uptick in ivory consumption by imposing export and import bans, stepping up anti-poaching patrols and launching information campaigns to educate Chinese consumers about the true cost of ivory.[82] Whether these measures will be enough to save elephants from extinction remains to be seen.

Musk

Musk comes from the musk deer (*Moschus moschiferus*), a small species of deer found in the Himalayas of India, Tibet and China and in the Altaic range near Lake Baikal. Produced only by the male deer, it is found in a small pocket close to the animal's navel. When fresh, musk is of a dark reddish-brown colour; once removed from the pod if becomes nearly black.

Musk was used in China from as early as 200AD and in India and Central Asia from around 500AD. It was first imported into Persia by the Parthians (247BC–224AD) and the Sasanians (224–651AD), serving as a diplomatic gift between kings, princes and caliphs. In 1001, for instance, the Karakhnid Nasir al-Haqq presented Mahmid of Ghazna with various exotic imports of the Turks, including 'ingots of precious metals from the mines, musk pods ... male horses, reddish-white female camels, attractive male and female slaves, white falcons, black fur [and]m carved khutu-ivory'.[83]

Following the Islamic conquest of Persia in the seventh century, musk was adopted across the Islamic World, serving as both an aromatic and a drug. Men rubbed it into their beards and moustaches. Women employed it as a cosmetic and to perfume sandals. Cooks used it as a food preservative. Physicians prescribed it to treat snake bites, heart palpitations and chronic headaches.

From the Arabic world, musk passed to medieval Europe, where it was used in pomanders to ward off the plague and sprinkled on gloves as a perfume. It was popular with Elizabeth I, who used it to scent writing paper and powder her face. It was also a favourite of Josephine Bonaparte, who used it in her bath. Starting in the reign of Charles I (1625–1649), musk was used in the anointing oil at the coronation of British kings, along with civet and ambergris.

In order to obtain musk, musk deer were hunted across Central Asia using dogs and snares. Several ninth-century Islamic writers (who had never seen the animal in person), suggested that deer excreted the musk naturally, rubbing against stones or stakes inserted in the ground by humans for this purpose. In reality the process was more brutal and required the capture and slaughter of many animals. Writing in the seventeenth century, Jean-Baptiste Tavernier described how musk deer would leave the forest each winter to feed upon corn and rice, whereupon, 'the Natives lay gins and snares ... to catch them as they go back: shooting some with Bows, and knocking others o' the heads'.[84] Two centuries later Colonel Frederick Markham gave a similar account of the collection of musk in northern India, describing how musk deer were 'hunted down with dogs' or snared in traps.[85]

Once sourced from the deer, musk was sold to travelling merchants by local collectors and exported across Eurasia. Some was carried overland by Sogdian traders (and later by other Turkic peoples). Some was transported by land to India, then shipped from Sind to the Arabian Peninsula. Some was shipped all the way from China. Consumers in the medieval Islamic world considered Tibetan musk to be of the highest quality because musk brought overland suffered less from the effects of moisture than musk transported on long sea voyages. Musk delivered in pods (made of the skin of the musk deer) was regarded as better than musk delivered in phials, since the latter were more susceptible to adulteration. Writing in the nineteenth century, Markham alleged that some of the pods brought to market 'were merely a piece of musk-deer skin filled with some substance, and tied up to resemble a musk-pod'.[86] Other common adulterants included tree sap, crushed lentils, acorn shells and goat's blood.

From the mid-nineteenth century the appeal of musk started to wane in the West, as consumer tastes shifted away from pungent animal perfumes and towards more delicate floral scents such as jasmine, lavender and rose. The creation of the first artificial musk in the laboratory further accelerated the move away from animal perfumes, offering a cheaper, synthetic alternative to the costly animal scent. By contrast, musk remained popular in China, where it continues to be used in traditional medicine as a treatment for poor blood circulation, spasms and cancer.

As the most expensive substance derived from an animal, musk has served for centuries as a coveted commodity. It contributed to the rise of trade emporia at Sogdiana, Bukhara and Samarkand and helped to forge commercial and cultural connections between Europe, Arabia and Central Asia. Like many animal products, musk has risen and fallen in popularity over time in response to shifting tastes and ethical considerations, while the means of harvesting it have changed; since 1958, musk has largely been extracted from semi-domesticated musk deer reared on farms in China, rather than by hunting and killing wild animals. The history of musk thus reveals the enduring appeal of some high-value animal products, but also the ways in which changing fashions, technological innovation and ecological concerns influenced their production and use.

Cochineal

Cochineal is a red dye, made from the crushed bodies of the cochineal bug (*Dactylopius coccus*). Cultivated in Mexico, it was consumed globally in Europe, the Middle East and China, becoming one of the most coveted animal products in the early modern world.

A carminic acid, cochineal is produced by the cochineal insect to deter predators such as ants, birds and armadillos. Only the females produce the dye; they spend their entire lives attached to the leaves of the nopal cactus, on which they lay their eggs. Though a wild species of the insect exists, the domesticated variety was believed to produce the best dye, forming the basis of the global trade in cochineal.

Cochineal was first cultivated in the Oaxaca region of Mexico. It was much prized by the Aztecs who collected cochineal as tribute from their subject peoples and used it to colour textiles, feathers, cosmetics and tamales; according to one Aztec document, the Zapotec peoples of central Oaxaca had to contribute twenty bags of grana cochinilla to the last Aztec Emperor, Moctezuma II, every three months. To cultivate the cochineal beetle, indigenous people in Oaxaca engaged in a complex and labour-intensive process that involved planting the cacti on which the insect lives, seeding each leaf with around fifty baby beetles, tending to the leaves regularly to eliminate predators and then carefully harvesting the mature beetles before the rainy season destroyed them. Once harvested, the beetles were killed by boiling them in water baths known as *temazcalli*, or by baking them in the sun. One pound of dye required the desiccation of 70,000 tiny insects – an indication of the huge amount of insect life and human labour that went into the production of cochineal.[87]

Following the Spanish conquest of Mexico in 1521, cochineal was exported to Europe, replacing madder (made from the root of a plant) and kermes (made from another species of beetle) as the red dye of choice for cushions, clothing and upholstery. Still produced in Oaxaca, cochineal was shipped from Veracruz to Seville and forwarded from there to major textile manufacturing centres in Segovia, Lyon, Rouen, Florence, Venice and Amsterdam. From Europe, cochineal-dyed textiles quickly acquired a global market, finding favour in the Middle East, India and China (accessed via the Spanish-governed Philippines). Reflecting on the dye's global significance, the Prussian polymath Alexander von Humboldt classified the cochineal beetle as 'an insect of the highest importance for the manufactures of Europe'.[88]

Due to its high value, cochineal became a target for biopiracy in the seventeenth and eighteenth centuries, as Spain's imperial rivals attempted to secure the insect for themselves. In the seventeenth century, English and Dutch privateers attacked Spanish galleons carrying cochineal (and other valued commodities) across the Atlantic. In the 1770s the French botanist Nicholas Joseph Thierry de Menonville smuggled live cochineal beetles out of Mexico and tried to cultivate them on the French colony of Saint Domingue, providing France with its own independent supply of the prized commodity. These ventures were unsuccessful, partly because of the strict controls imposed by the Spanish government and partly because of the delicacy of the cochineal beetle, which required assiduous care and the right climatic conditions to survive. In the nineteenth century, however, cochineal was successfully transplanted outside of Oaxaca, with introductions of the beetles to Guatemala and, later the Canary Islands; by 1870 the latter had become the main global producer of the dye.[89]

From the 1850s (and especially the 1870s) the demand for cochineal began to subside as new synthetic dyes appeared on the market. Cheaper than cochineal, aniline dyes (made from coal tar) largely replaced natural dyes in textile manufacturing, pushing many producers out of business. Changes in fashion also reduced the popularity of cochineal, as taste shifted towards more muted colours. Today cochineal is used primarily as a food colouring, appearing in products as diverse as yogurts, sausages, ice cream, cough syrup and lipstick. Peru is the world's primary exporter of the substance, producing 85% of the cochineal on the market.[90]

The story of cochineal highlights the role of geopolitics, technological change and shifting tastes in the global reach and popularity of an animal commodity. The product of Amerindian expertise, it revolutionised world textile production in the sixteenth century, spurred imperial competition in the seventeenth and eighteenth centuries and faded from view in the nineteenth century in the face of chemical innovation. Like pork, ivory and musk, its use was shaped by global exchanges, shifts in taste and technological change.

Conclusion

Animals have been consumed throughout human history. Hunted, herded and, from the twentieth century, intensively farmed, they have provided humans with meat, clothing, fuel and fertiliser. Some animal products have been traded on a global scale, forging connections between distant societies.

Various factors have shaped the consumption of animal commodities, from technological innovation to ethical considerations. Global exploration has improved access to furs, feathers and ivory. Colonisation has opened new land for raising livestock. New machinery has speeded up the transportation and processing of animal products, while domestication and selective breeding have reshaped animal bodies for human benefit. Cultural and ethical factors have also influenced the demand for certain animal commodities, confining the wearing of pearls and fur to specific social groups and pushing some consumers to renounce meat or egret plumes. Studies of animal commodities reveal temporal and geographical variations in the use of animal products and highlight the mechanisms through which local products became regional or global. By placing commodities within their wider historical contexts, historians can help us to understand why tapirs' hooves came to be viewed as a cure for epilepsy in colonial Spanish America, why there was a boom in ostrich farming in 1880s South Africa and how Holstein cows from Germany came to dominate dairy herds in twenty-first-century California.

Notes

1 Mark Kurlansky, *Cod: A Biography of the Fish that Changed the World* (London: Vintage Books, 1999).
2 Sarah Abreyava Stein, *Plumes: Ostrich Feathers, Jews and a Lost World of Global Commerce* (New Haven: Yale University Press, 2008).
3 Peter Borschberg, 'The Euro-Asian Trade in Bezoar Stones (approx. 1500 to 1700)', in Michael North (ed.), *Artistic and Cultural Exchanges between Europe and Asia, 1400–1900* (Farnham: Ashgate, 2010), pp. 29–43.
4 Irina Podgorny, 'The Elk, the Ass, the Tapir, their Hooves and the Falling Sickness: A Story of Substitution and Animal Medical Substances', *Journal of Global History* 13 (2018), pp. 46–68.
5 Bernabé Cobo, *Historia del Nuevo Mundo* (Sevilla: Imprenta de E. Rasco, 1895), Vol. II, p. 335.
6 Marcia Stephenson, 'From Marvelous Antidote to the Poison of Idolatry: The Transatlantic Role of Andean Bezoar Stones during the Late Sixteenth and Early Seventeenth Centuries', *Hispanic American Historical Review* 90:1 (2010), pp. 3–39.
7 Liz P. Chee, *Mao's Bestiary: Medicinal Animals and Modern China* (Durham, NC: Duke University Press, 2021), pp. 139–160.
8 'No Decline in Endangered Species Trading', *The Times*, 5 June 1980.
9 Jonathan Schlesinger, *A World Trimmed with Fur: Wild Things, Pristine Places and the Natural Fringes of Qing Rule* (Stanford: Stanford University Press, 2017), p. 25.
10 Helen Cowie, *Victims of Fashion: Animal Commodities in Victorian Britain* (Cambridge: Cambridge University Press, 2022), pp. 55–86.
11 Robin Doughty, *Feather Fashions and Bird Preservation: A Study in Nature Protection* (Berkeley: University of California Press, 1975).
12 Charles Ledger, 'The Alpaca', in George Bennett, MD, FLS, *The Third Annual Report of the Acclimatisation Society of New South Wales* (Sydney, 1864), pp. 92–93.
13 Gregory T. Cushman, *Guano and the Opening of the Pacific World: A Global Ecological History* (Cambridge: Cambridge University Press, 2013).
14 John F. Richards, *The World Hunt: An Environmental History of the Commodification of Animals* (Berkeley: University of California Press, 2014), p. 145.
15 Briton Cooper Busch, *The War Against the Seals: A History of the North American Seal Fishery* (Montreal: McGill University Press, 1985), p. 182.
16 John MacKenzie, *The Empire of Nature: Hunting, Conservation and British Imperialism* (Manchester, Manchester University Press, 1988); Emma Griffin, *Blood Sport: Hunting in Britain since 1066* (New Haven: Yale, 2007).
17 Erica Hill, 'Pre-Domestication: Zooarchaeology', in Brett Mizelle, André Krebber, Mieke Roscher and Aline Steinbrecher (eds), *Handbook for Historical Animal Studies* (Berlin: De Gruyter, 2021), pp. 21–36.
18 Marcelo Sánchez-Villagra, *The Process of Animal Domestication* (Princeton: Princeton University Press, 2022, pp. 1–35.
19 Bin Yang, 'The Rise and Fall of Cowrie Shells: The Asian Story', *Journal of World History* 22 (2011), pp. 1–25.
20 Molly Warsh, *American Baroque: Pearls and the Nature of Empire, 1492–1700* (Chapel Hill: University of North Carolina Press, 2018).
21 Janet Martin, *Treasure of the Land of Darkness: The Fur Trade and its Significance for Medieval Russia* (Cambridge: Cambridge University Press, 1986), p. 104.

22 Jonas Kranzer, 'Tickling and clicking the ivories: The metamorphosis of a global commodity in the nineteenth century', in Bernd-Stefan Grewe and Karin Hofmeester (eds), *Luxury in Global Perspective: Objects and Practices 1600–2000* (Cambridge: Cambridge University Press, 2016), pp. 242–262.
23 Chris Otter, 'Eating Animals', in Philip Howell and Hilda Kean (eds), *The Routledge Companion to Animal–Human History* (London: Routledge, 2018), p. 480.
24 Pablo José de Arriaga, *Extirpación de la Idolatría del Perú* (Lima: Geronymo de Contreras, 1621), p. 25.
25 'Country House', *The Bazaar*, 7 January 1903, p. 71.
26 Alfred Crosby, *The Columbian Exchange: Biological and Cultural Consequences of 1492* (Westport: Greenwood Press, 1972), pp. 88 and 92.
27 Shepard Krech III, *The Ecological Indian* (New York: W.W. Norton, 1999), pp. 173–209.
28 Sharika Crawford, *The Last Turtlemen of the Caribbean: Waterscapes of Labor, Conservation and Boundary Making* (Chapel Hill: University of North Carolina Press, 2020), pp. 15–38.
29 Abdul Sheriff, *Slaves, Spices and Ivory in Zanzibar* (Athens: Ohio University Press, 1987), pp. 77–115.
30 Kurlansky, *Cod*, pp. 78–90.
31 David Nibert, *Animal Oppression and Human Violence: Domesecration, Capitalism and Global Conflict* (New York: Columbia University Press, 2013), pp. 92–170.
32 Carolyne Larson (ed.), *The Conquest of the Desert: Argentina's Indigenous Peoples and the Battle for History* (Albuquerque: University of New Mexico Press, 2020), p. 9. These are the official figures from the Argentine army and are likely a significant underestimate.
33 John Soluri, *Creatures of Fashion: Animals, Global Markets and the Transformation of Patagonia* (Chapel Hill: Duke University Press, 2024), pp. 47–74.
34 Virginia DeJohn Anderson, *Creatures of Empire: How Domestic Animals Transformed Early America.* (Oxford: Oxford University Press, 2004).
35 Cowie, *Victims of Fashion*, pp. 38, 144 and 59–60.
36 Joshua Specht, *Red Meat Republic: A Hoof-to-Table History of How Beef Changed America* (Princeton: Princeton University Press, 2019), pp. 118–173.
37 Rebecca Woods, 'From Colonial Animal to Imperial Edible: Building an Empire of Sheep in New Zealand, ca.1880–1900', *Comparative Studies of South Asia, Africa and the Middle East* 35:1 (2015), pp. 117–136.
38 Richard Bulliet, *Hunters, Herders and Hamburgers* (New York: Columbia University Press, 2005), pp. 1–35.
39 Bulliet, *Hunters, Herders and Hamburgers*, pp. 101–142.
40 Abel Alves, 'Domestication: Co-evolution', in Brett Mizelle, André Krebber, Mieke Roscher and Aline Steinbrecher (eds), *Handbook for Historical Animal Studies* (Berlin: De Gruyter, 2021), pp. 37–51. Conversely, the biology of some species rendered them unsuitable for domestication, whether as a result of their aggressiveness (zebras, hippopotami), their nervousness and territoriality (deer, gazelles), their slow rate of reproduction (elephants) or their specialist diets (koalas).
41 Juliet Clutton-Brock, *A Natural History of Domesticated Mammals* (London: British Museum, 1987), pp. 23–24.
42 Harriet Ritvo, *The Animal Estate: The English and Other Creatures in the Victorian Age* (Cambridge, MA: Harvard University Press, 1987), pp. 45–81.

43 Rebecca Woods, *The Herds Shot Around the World: Native Breeds and the British Empire, 1800–1900* (Chapel Hill: University of North Carolina Press, 2017), pp. 140–164.
44 Otter, 'Eating Animals', p. 480.
45 Karl Steel, *How to Make a Human: Animals and Violence in the Middle Ages* (Columbus: Ohio State University Press, 2011), pp. 67–91.
46 Antonio de León Pinelo, *El Paraíso en el Nuevo Mundo* (Lima: Imprenta Torres Aguirre, 1943), Vol. II, p. 19.
47 'A Voracious Beast', *Bristol Mercury*, 19 September 1835.
48 Kari Weil, *Precarious Partners: Horses and their Humans in Nineteenth-Century France* (Chicago: University of Chicago Press, 2020), pp. 84–102.
49 Otter, 'Eating Animals', pp. 484–486.
50 'French Dietetics', *The Belfast Newsletter*, 23 February 1856.
51 'Horsemeat Scandal: The Essential Guide', *The Guardian*, 15 February 2013.
52 Cowie, *Victims of Fashion*, pp. 188–196.
53 Martha Chaiklin, 'Tortoiseshell in Early Modern Japan', in Bernd-Stefan Grewe and Karin Hofmeester (eds), *Luxury in Global Perspective: Objects and Practices, 1600–2000* (Cambridge: Cambridge University Press, 2016), pp. 231–2.
54 Schlesinger, *A World Trimmed with Fur*, p. 28.
55 Bernabe Cobo, *Historia del Nuevo Mundo* (Seville, Imprenta de E. Rasco, 1895), Vol. IV, p. 205.
56 'Colobus Monkeys from Kenya', *The Times*, 27 June 1925.
57 'Investigation of Fur-Seal Industry of Alaska', 8 June 1912, p. 1011, cited in William T. Hornaday, *Scrapbook Collection on the History of Wild Life Protection and Extermination*, Vol.5, Wildlife Conservation Society Archives Collection, 1007-04-05-000-a.
58 'The London Fur Trade', *Chambers' Journal of Popular Literature, Science and Art*, 6 October 1894, p. 626.
59 Shuk-Wah Poon, 'Dogs and British Colonialism', *The Journal of Imperial and Commonwealth History* 42:2 (2014), pp. 308–328.
60 Lettice McNaughten, 'Ways of Helping', *The Animals' Friend*, 1910, p. 115.
61 McNaughten, 'Ways of Helping', p. 115.
62 James Gregory, *Of Victorians and Vegetarians: The Vegetarian Movement in Nineteenth-Century Britain* (London: Tauris, 2007), pp. 88–110.
63 Henry Salt, *Seventy Years Among Savages*, (London: George Allen & Unwin, 1921), p. 64.
64 'The Bird of Paradise', *RSPB Pamphlet Number 20*, 1895, p. 2.
65 'The Fur Industry', *The Animals' Friend*, September 1897, p. 244.
66 Specht, *Red Meat Republic*, pp. 218–246.
67 Jonathan Saha, 'Milk to Mandalay: Dairy Consumption, Animal History and the Political Geography of Colonial Burma', *Journal of Historical Geography* 54 (2016), pp. 1–12.
68 Otter, 'Eating Animals', p. 476.
69 Hannah Ritchie et al., 'Population Growth' (2023), retrieved from https://ourworldindata.org/world-population-growth.
70 Sam White, 'From Globalised Pig Breeds to Capitalist Pigs: A Study in Animal Cultures and Evolutionary History', *Environmental History* 16 (2011), pp. 95–98.
71 White, 'From Globalised Pig Breeds to Capitalist Pigs', pp. 98–109.
72 'A Pretty Couple', *The Graphic*, 21 December 1901.
73 Otter, 'Eating Animals', p. 478.
74 Brett Mizelle, *Pig* (London: Reaktion, 2011), pp. 41–54.

75 Otter, 'Eating Animals', 480.
76 Martina Siebert, 'Reforming the Humble Pig: Pigs, Pork and Contemporary China', in Roel Sterckx, Martina Siebert and Dagmar Schäfer (eds), *Animals Through Chinese History: Earliest Times to 1911* (Cambridge: Cambridge University Press, 2019), pp. 233–243.
77 Cowie, *Victims of Fashion*, pp. 90–94.
78 'The African Elephant', *The Anti-Slavery Reporter*, January 1879, p. 139.
79 'Editorial', *The Times of India*, 13 June 1882.
80 'Indian and African Elephants', *The Graphic*, 28 September 1878.
81 Cowie, *Victims of Fashion*, pp. 97–125.
82 Cowie, *Victims of Fashion*, pp. 254–256.
83 Anya King, *Scent from the Garden of Paradise: Musk and the Medieval Islamic World* (Leiden: Brill, 2017).
84 Jean-Baptiste Tavernier, *The Six Voyages of John Baptista Tavernier, Baron of Aubonne through Turky, into Persia and the East-Indies, for the Space of Forty Years* (trans. Daniel Cox) (London: William Godbid, Robert Littlebury and Moses Pitt, 1677), Travels in India, The Second Book, p. 153.
85 Frederick Markham, *Shooting in the Himalayas: A Journal of Sporting Adventures and Travel* (London: Richard Bentley, 1854), pp. 94–96.
86 Markham, *Shooting in the Himalayas*, pp. 98–100.
87 Carlos Marichal Salinas, 'Mexican Cochineal, Local Technologies and the Rise of Global Trade from the Sixteenth to the Nineteenth Centuries', in M. Perez Garcia and L. de Sousa (eds), *Global History and New Polycentric Approaches* (New York: Palgrave Macmillan, 2018), pp. 255–259.
88 Alexander von Humboldt, *Essai Politique sur le Royaume de la Nouvelle-Espagne* (Paris: J.H. Stône, 1811), Vol. III, p. 245.
89 Edward Melillo, 'Global Entomologies: Insects, Empires and the "Synthetic Age" in World History', *Past and Present* 223 (2014), pp. 255–257.
90 Melillo, 'Global Entomologies', p. 266.

Further Reading

Abreyava Stein, Sarah, *Plumes: Ostrich Feathers, Jews and a Lost World of Global Commerce* (New Haven: Yale University Press, 2008)
Alves, Abel, 'Domestication: Co-evolution', in Brett Mizelle, André Krebber, Mieke Roscher and Aline Steinbrecher (eds), *Handbook for Historical Animal Studies* (Berlin: De Gruyter, 2021), pp. 37–51
Bulliet, Richard, *Hunters, Herders and Hamburgers* (New York: Columbia University Press, 2005)
Chaiklin, Martha, 'Imports and Autarky: Tortoiseshell in Early-Modern Japan', in Bernd-Stefan Grewe, and Karin Hofmeester (eds), *Luxury in Global Perspective: Objects and Practices 1600–2000* (Cambridge: Cambridge University Press, 2016), pp. 218–241
Chee, Liz P., *Mao's Bestiary: Medicinal Animals and Modern China* (Durham: Duke University Press, 2021)
Cowie, Helen, *Victims of Fashion: Animal Commodities in Victorian Britain* (Cambridge: Cambridge University Press, 2022)
Crawford, Sharika, *The Last Turtlemen of the Caribbean: Waterscapes of Labor, Conservation and Boundary Making* (Chapel Hill: University of North Carolina Press, 2020)

Crosby, Alfred, *The Columbian Exchange: Biological and Cultural Consequences of 1492* (Westport: Greenwood Press, 1972)

Cushman, Gregory, *Guano and the Opening of the Pacific World: A Global Ecological History* (Cambridge: Cambridge University Press, 2013)

DeJohn Anderson, Virginia, *Creatures of Empire: How Domestic Animals Transformed Early America*. (Oxford: Oxford University Press, 2004)

King, Anya, *Scent from the Garden of Paradise: Musk and the Medieval Islamic World* (Leiden: Brill, 2017)

Kurlansky, Mark, *Cod: A Biography of the Fish that Changed the World* (London: Vintage Books, 1999)

Melillo, Edward, 'Global Entomologies: Insects, Empires and the "Synthetic Age" in World History', *Past and Present* 223 (2014), pp. 233–270

Mizelle, Brett, *Pig* (London: Reaktion, 2011)

Nibert, David, *Animal Oppression and Human Violence: Domesecration, Capitalism and Global Conflict* (New York: Columbia University Press, 2013)

Otter, Chris, 'Eating Animals', in Philip Howell and Hilda Kean (eds), *The Routledge Companion to Animal–Human History* (London: Routledge, 2018), pp. 474–498

Podgorny, Irina, 'The Elk, the Ass, the Tapir, their Hooves and the Falling Sickness: A Story of Substitution and Animal Medical Substances', *Journal of Global History* 13 (2018), pp. 46–68

Poon, Shuk-Wah, 'Dogs and British Colonialism', *The Journal of Imperial and Commonwealth History* 42:2 (2014), pp. 308–328

Richards, John, *The World Hunt: An Environmental History of the Commodification of Animals* (Berkeley: University of California Press, 2014)

Saha, Jonathan, 'Milk to Mandalay: Dairy Consumption, Animal History and the Political Geography of Colonial Burma', *Journal of Historical Geography* 54 (2016), pp. 1–12

Schlesinger, Johnathan, *A World Trimmed with Fur: Wild Things, Pristine Places and the Natural Fringes of Qing Rule* (Stanford: Stanford University Press, 2017)

Sheriff, Abdul, *Slaves, Spices and Ivory in Zanzibar* (Athens: Ohio University Press, 1987)

Siebert, Martina, 'Reforming the Humble Pig: Pigs, Pork and Contemporary China', in Roel Sterckx, Martina Siebert, and Dagmar Schäfer (eds), *Animals Through Chinese History: Earliest Times to 1911* (Cambridge: Cambridge University Press, 2019), pp. 233–243

Specht, Joshua, *Red Meat Republic: A Hoof-to-Table History of How Beef Changed America* (Princeton: Princeton University Press, 2019)

Warsh, Molly, *American Baroque: Pearls and the Nature of Empire, 1492–1700* (Chapel Hill: University of North Carolina Press, 2018)

Weil, Kari, 'Let Them Eat Horse', in Weil, Kari, *Precarious Partners: Horses and their Humans in Nineteenth-Century France* (Chicago: University of Chicago Press, 2020), pp. 84–102.

White, Sam, 'From Globalised Pig Breeds to Capitalist Pigs: A Study in Animal Cultures and Evolutionary History', *Environmental History* 16 (2011), pp. 94–120

Woods, Rebecca, *The Herds Shot Around the World: Native Breeds and the British Empire, 1800–1900* (Chapel Hill: University of North Carolina Press, 2017)

2   Labour

Animals have served humans as more than just food on the hoof. From as early as the Pleistocene era, when dogs were first domesticated, they have provided them with a vital source of labour, helping them to hunt prey, plough fields and transport goods and people from place to place. Animals have also assisted humans in waging war, furnishing armies with vital logistical support and carrying soldiers into battle. Though the scope, volume and nature of animal labour has changed over time there is little doubt that working animals have played a pivotal role in world history, facilitating global trade and even contributing to the rise and fall of civilisations.

Given the importance of animals as workers, scholars have paid comparatively little attention to this aspect of animal history, perhaps because working animals are (with the possible exception of military animals) less glamorous than pets or zoo animals and, in consequence, have generated less documentation. In recent decades, however, this situation has started to change. Focusing on a range of working animals, from oxen and camels in Ottoman Egypt to elephants on the timber plantations of Burma, historians have begun to explore the dynamics of human–animal relationships within a labouring context and to show how animal power shaped commerce, social hierarchies and military tactics. They have also posed questions about the ethics of animal labour, and the status assigned to working beasts. Are labouring animals partners or slaves? Is animal labour cruel, or can it benefit animals as well as humans? To what extent has the use (or non-use) of animal labour influenced social and cultural development? Drawing on this growing body of scholarship, this chapter outlines the key debates surrounding animal labour and assesses how it has changed over time.

Animal Power

Animal labour has played a crucial role in the development of human societies across the globe. Before the invention of the steam engine, animals were the main source of power for all kinds of tasks, assisting in the transportation

DOI: 10.4324/9781003181996-3

of commodities and people and the operation of machinery. Though the advent of steam and later electric power has since reduced the number of labouring animals in many societies, animals continue to work for humans in a variety of different guises, serving as pest controllers, protectors of people and property and sources of emotional support.

Of all the contributions made by animals to human society, one of the most important has been their role in the development of human agriculture. Before the domestication of cattle, humans tilled the soil themselves, using simple hoes or digging sticks. From around 2000BC, however, animal power was enlisted for the ploughing of fields, enabling humans to significantly increase the yield of crops such as wheat, rice and barley. The Egyptians used long-horn cattle for ploughing from at least 1300BC, while the Romans used oxen (castrated bulls) to plough fields across southern Europe, substantially increasing wheat yields in the region and facilitating a more intensive form of agriculture. Other species have also performed agricultural labour, most notably horses. From the eighteenth century, the latter were bred specifically for ploughing, giving rise to iconic breeds such as the Shire horse and the Suffolk Punch.

A second area in which animal energy played a vital role was in powering heavy machinery. In Ottoman Egypt oxen powered waterwheels to irrigate the fields that bordered the Nile.[1] In early modern England small dogs with short legs known as turnspits ran inside wheels in the kitchen to roast beef on a spit. In nineteenth-century East Africa camels were employed in the extraction of coconut oil. As well as producing goods for local consumption, animal power was also essential to the processing of several global commodities, often complementing the work of enslaved and indentured human labourers. Oxen and horses powered sugar mills in Brazil and the Caribbean, moving cylindrical rollers to extract the juice from the cane.[2] Tame elephants performed crucial labour on colonial timber plantations in nineteenth-century Burma, transporting and stacking teak for use by the British navy.[3] Without the miles walked by animals in waterwheels and around millstones, fields could not have been irrigated, grain could not have been ground and prized commodities could not have been processed in the volumes required for regional and global exchange.

A third major area in which animal power has been exploited by humans has been in the transportation of goods. At the local level, horses, donkeys and mules have carried grain, timber, wool and manufactures from the field or the workshop to the marketplace, serving as vital cogs in a range of industries. At a regional level, animal porterage has facilitated the movement of commodities across continents, giving rise to complex trade routes and cross-cultural interactions. Camels, for instance, carried salt and gold across the Sahara and silk across Central Asia. Llamas lugged silver from Potosí to the Pacific, while mules carried cochineal from Oaxaca to the Caribbean port of Veracruz. The number of animals involved in these operations was large, attesting their importance to the global supply chain. In the case of llamas, Peruvian Jesuit Antonio León Pinelo reported that by the early seventeenth century, some eight thousand camelids were employed in transporting silver from the mines of Potosí in Upper Peru (modern-day Bolivia), playing a crucial role in the colonial economy.[4] Two centuries later British wool merchant Charles Ledger estimated that around 600,000 llamas transported 'seven-eighths of the products of the country [Peru] to the coast for export, [including] wool, copper in bar and ore, tin in bar and ore' as well as carrying 'the silver ores from mine to amalgamating establishments' and bringing 'provisions … into all the cities, towns and hamlets of the interior'.[5] Animal caravans were therefore fundamental to the internal and transcontinental trade in a range of important commodities, from silk to silver.

Crucially, some animals – most notably horses and donkeys – have been used for human transportation. Horses, camels and reindeer have pulled humans in chariots and sledges, carried them in wagons and stagecoaches and, in the nineteenth century, hauled them through city streets in cabs and omnibuses. Horses, mules, donkeys, camels, and elephants have also been ridden, providing transportation in peace and cavalry in war. By carrying humans over distances they could not comfortably have walked, animals have enabled people to commute to work, travel between cities and colonise distant regions, revolutionising communications, lifestyles and settlement patterns. Horses, moreover, have played a key role in hunting and, later, ranching, transforming human relationships with wild animals. The introduction of horses into the North American plains, for instance, had a massive impact on indigenous peoples such as the Sioux, who were able to hunt bison much more efficiently than they had previously been able to do on foot. This, in turn, allowed Native Americans to engage in hunting throughout the year, abandoning agricultural farming and embracing a nomadic existence in search of migratory bison herds.

Finally, as well as providing humans with energy to move people and goods, animals have assisted humans in two additional ways: as protectors against other, less welcome, species and as extensions of the sensorially impoverished, vulnerable or disabled human body. In the former case, a variety of animals have combated predators, pathogens and vermin by hunting and consuming pests and disposing of human and animal waste. Dogs, for instance, have been used to defend livestock from wolves and property from human thieves. Cats and ferrets have served humans by hunting rats and mice, which would otherwise have consumed grain intended for human sustenance, while scavengers have cleansed city streets of dead animal carcasses, removing a potential source of infection from human settlements. In the Ethiopian town of Harar, hyenas gather each night to consume rotting meat, removing a health hazard from the streets.[6] Several less conventional animals have also been recruited to shield humans from bespoke biological and man-made threats, taking advantage of some niche anatomical traits or nutritional preferences. In 1964, for example, marine biologists and flood control officials recruited manatees to clear invasive hyacinths from Florida's canals, deploying the amphibious herbivores as 'mammalian mowing machines'.[7] In 1897, when 'a horde of hairy yellow caterpillars infested the linden trees' at Philadelphia Zoological Gardens, Head Keeper Manley enlisted the services of the institution's giant anteater, taking the animal out 'three times a day' on 'a collar and a long cord' and letting it scoop up the creatures 'by the dozen' with its 'long, sticky tongue'.[8]

Animal assistance covers a wide range of activities, but has included compensating for physical or sensory impairments in humans, performing complex tasks and providing psychological support. Dogs, for example, have worked with the police in tracking criminals and detecting narcotics. They have also functioned as guides for the deaf and the visually impaired. Pigs have been used to sniff out truffles in France and Italy – a practice dating back to the Roman Empire. Capuchin monkeys have been trained to assist quadriplegics, performing tasks such as 'operating lights; taking a sandwich from a refrigerator and putting it on a tray; putting a thermos in a holder on a tray, opening it and inserting a straw'.[9] Cats, rabbits and llamas have served humans as therapy animals, offering companionship to schoolchildren, hospital patients and nursing home residents. Performing these more complex types of work requires keenly developed senses, specific physical characteristics and, in almost all cases, intensive training, as well as a considerable dose of animal aptitude and human ingenuity. While sometimes coerced, this kind of labour is often the result of a close partnership between human and animal, which, though engineered for anthropocentric reasons, can have positive results for both parties.

The Impacts of Animal Labour

Animal labour has thus enabled humans to exploit the strength, endurance and sensory acuity of other species to perform a wide array of tasks, from transporting goods to herding sheep. How, though, has animal labour influenced the development of human societies on a global scale? In what ways has the nature and volume of animal labour changed over time? What have been the ethical implications of using non-humans to work for human benefit? These questions have preoccupied historians and continue to shape our interactions with working animals to this day.

Viewed from a world history perspective, the presence or absence of labouring animals in specific cultures has had a direct impact on their development, shaping their demographics, social organisation and even the layout of their settlements. In Pre-Columbian America, for example, the lack of domesticated draught animals (with the exception of the llama), meant that crucial jobs such as ploughing and porterage all had to be done by humans, while hunting and warfare also had to be conducted on foot. Civilisations such as the Maya, the Aztecs and the Inca overcame these challenges through the effective mobilisation of human labour; the Inca mustered thousands of labourers each year to construct palaces, maintain agricultural terraces and serve in the imperial army. The absence of working animals has, however, left its mark on their Pre-Columbian societies in ways that are still evident in their architecture. The streets of the Inca capital, Cusco, designed to accommodate only human and camelid traffic, were unsuitable for horses or wheeled vehicles (which the Inca did not have).

The impacts of animal labour have also been profound in Asia and North Africa, where the domestication of the camel as a beast of burden connected distant societies and left a visible mark on the landscape. Capable of coping with losing up to a quarter of their body weight to dehydration, camels were able to traverse otherwise inhospitable deserts, carrying goods over long distances. In the Arabian Peninsula and North Africa, camel trains ferried salt, frankincense and gold across the Arabian desert and the Sahara, shaping Bedouin culture and giving rise to important cities such as Palmyra, Petra and Timbuktu, which emerged as entrepôts along key trading routes. In Asia camels transported coveted silk and spices along the Silk Road, linking silk producers in China with consumers as distant as Istanbul and Venice. While wheeled vehicles existed in the Middle East prior to *c*.200AD, moreover, historian Richard Bulliet has argued that the latter fell into disuse in those regions where the camel predominated as a beast of burden, owing to the greater efficiency of camel-based transport in sandy, arid environments. This had significant consequences for the development of road systems in the Middle East and North Africa, and, as in the Americas, for the layout of cities: 'As late as 1845 the width of a major new street in Cairo was determined by measuring the combined width of two loaded camels.'[10]

In more recent centuries, humans have consciously transplanted valued animal species to new settings and continents for the specific purposes of performing labour and opening territory to settlement. In the late nineteenth century, British colonists imported camels from Baluchistan in modern-day Pakistan to Northern and Western Australia to facilitate the exploitation of a dry, desolate region with no native draught animals and a climate unsuited to the horse. The scheme initially failed, with many of the earliest camels succumbing to disease (predominantly mange) shortly after importation. The recruitment of Afghan camel handlers, however, and the institution of quarantine measures, overcame this challenge, giving rise to a highly successful acclimatisation programme. By 1894, there were thought to be nearly 10,000 camels at work in Australia, where they were used to carry supplies to pastoralists, to assist in gold mining operations and to transport well-sinking equipment to waterless regions.[11] One camel had reportedly 'been ridden on police service in South Australia' covering 'over 100 miles in one day'; a second, Snake, covered the 160 miles between Beltana and Port Augusta in just 24 hours.[12]

Another similar – though unsuccessful – animal training scheme occurred in the Belgian Congo, where efforts were made to convert elephants into beasts of burden. The Belgian colonists first attempted to acclimatise Asian elephants in the region, importing four of the animals from Mumbai in 1879. When this experiment failed, they turned their attention to indigenous African elephants, catching the animals as adolescents and training them to pull carts, carry timber and plough fields – valuable skills in a region where diseases spread by the tsetse fly quickly killed off imported cattle and horses. The enterprise had achieved some successes by 1928, when 'Four elephants [were] employed on the mission cultivations at Buta, two by Mr de Steenhault de Waerbeke, a planter at Dembea, and others on a cotton farm at Bambessa'.[13] Ultimately, however, the costs of catching, feeding and managing captive elephants proved too great, bringing the domestication scheme to an end.

If the use of animals for work has differed between places, it has also evolved over time, with significant changes in the volume and nature of animal labour. Taking the long view, the use of animal power broadly followed an upward trajectory until the middle of the nineteenth century, as human populations grew and the demand for food and transportation increased. From around 1850, however, that usage started to diminish, as steam and, from around 1900, electricity, began to power transport and machinery. While these generalisations capture the overall global trend in the deployment of animal labour, however, both processes have been slow and uneven, happening at a different pace in different parts of the globe. The expansion of animal labour, for instance, came late to the Americas and Australia, which lacked large herbivores suitable for ploughing or riding until contact with Europeans in the sixteenth and eighteenth centuries respectively. The virtual disappearance of draft animals from European and North American cities and fields following mechanisation, meanwhile, has not been replicated with equal speed in Asia, Latin America or Sub-Saharan Africa, where animal power continues to play a crucial role in trade and agriculture. In 2021 there were an estimated 70 million draught animals in India, ploughing approximately 65% of the country's cultivated land.[14] In Cuba, the economic decline triggered by the collapse of the Soviet Union in 1989 caused a reduction in imports of fossil fuels and automotive vehicles, leading to greater reliance on oxen, which increased in number from 163,000 in 1990 to nearly 400,000 by 2000.[15] Broad global trajectories for the rise and decline of animal labour can thus obscure significant local and regional variations.

A second feature that has marked animal labour over time has been a move towards increasing specialisation, as non-human workers were bred or trained to perform specific tasks. From the seventeenth century, distinct breeds of horse started to emerge, designed especially for agricultural work (the Suffolk Punch) or for pulling heavy vehicles (the Shire horse). In Ottoman Turkey Anatolian nomads bred dromedaries with Bactrian camels to create a physically stronger hybrid, better able to serve as beasts of burden in interregional trade.[16] Since the Roman era, meanwhile, humans have bred dogs to create animals more suited for hunting (greyhounds), retrieving prey (spaniels), catching rats (terriers), fighting bulls (bulldogs) or cooking meat on a spit (turnspits). In the early twentieth century this was followed by the introduction of the first police dogs in Britain, France and the USA, and the first efforts to train guide dogs for the visually impaired.[17] The expectations of working animals have thus changed over the centuries, with increased emphasis on bespoke qualities and skillsets. Attitudes towards animal labour have also been influenced by a range of other factors, including concerns about public health and evolving conceptions of animal welfare.

These global developments are nicely encapsulated in the experience of working horses in nineteenth- and twentieth-century North America. Always important in the USA as beasts of burden, horses experienced a massive spike in demand from the middle of the nineteenth century as cities such as New York, Boston and Chicago expanded. They played a vital role in human transportation, the distribution of goods and the powering of heavy machinery and had a huge impact on city life, whether through their requirements for food and stabling, their production of manure (valued in the nineteenth century as a fertiliser), their visible suffering on the city streets (which the American Society for the Prevention of Cruelty to Animals worked to mitigate) or their post-mortem contributions to the urban economy as leather, glue, phosphate (used in matches) or pet food. As in other parts of the world, different types of horse were bred and trained for different industries, reflecting the broader trend towards specialisation. Breweries preferred Clydesdales, primarily for their looks. Furniture movers selected slow horses, less likely to bolt while carrying fragile items. Undertakers enlisted black horses, for aesthetic reasons.[18]

From the 1870s, the number of horses working in American cities began to fall as their jobs were taken over by steam-powered machinery, electrified tram lines and, by the early 1900s, the motor car. As Clay McShane and Joel Tarr emphasise, however, this transition was gradual, occurring in different industries at different times and often involving a significant changeover period, during which animal and steam power operated side-by-side. Horses, for instance, pulled the new steam-powered fire engines in cities such as Boston and New York, while pit ponies laboured to mine the coal needed to fuel new steam-based machinery. Equine power also continued to be used for localised deliveries well into the twentieth century, competing success-fully with gasoline-fuelled vehicles. The decline of the urban horse was, therefore, a protracted process, and the horse's demise as a source of power by no means a certainty; as late as 1900 there were an estimated three million urban horses in the United States, despite the increasingly widespread use of fossil fuels.[19]

The Ethics of Animal Labour

This brings us nicely to a second important area of discussion: the ethics of animal labour and the treatment accorded to working animals. How are working animals perceived by humans? Does labour confer respect, or merely suffering? Are working animals exploited slaves or valued partners? Does animal labour increase or decrease human exploitation? The answers to these questions are complex and suggest that animal labour can have both positive and negative impacts on human–animal relations.

Given its anthropocentric focus, it is easy to view animal labour as cruel and exploitative. Animals do not usually have a choice about whether they wish to work for the benefit of humans, so their labour is, in most cases, involuntary. Cruelty has also been endemic in the treatment of working ani-mals, which are often viewed as 'living machines' to be worked as hard as possible for maximum profits.[20] In 1841, for instance, the Royal Society for the Prevention of Cruelty to Animals (RSPCA) prosecuted London omni-bus driver Thomas Horsoy for working a horse with 'a wound on the hip, another on the side, and a large wound on the chest', the latter apparently caused by the collar chafing against its neck.[21] Four years later, the RSPCA secured a conviction against a man named William Peacock for 'driving a donkey and hitting it with a thick stick ... in the most cruel manner' to make it go faster.[22] To some extent, therefore, animal labour can be seen as just another way in which humans have exploited animals, making them do dif-ficult, exhausting jobs in unpleasant conditions.

In certain circumstances, however, it is possible to view animal labour as a constructive partnership between humans and animals, in which the latter earn respect for their work and exert a degree of agency over their working patterns. If assigned to perform types of work that align closely with their natural behaviours, for example, some animals may enjoy working and gain a sense of satisfaction from the successful completion of a task. Buster, the star performer of dog trainer, Rennie Renfro (who supplied canine actors to Hollywood in the 1930s), reportedly 'jump[ed] into [his] arms for a kiss' after doing 'a good scene well', but skulked 'off in a dark corner' if he had been 'a "flop"'.[23] In other cases working animals can forge a close bond with their human handlers, who may, over time, learn to recognise the subtle cues indi-cating the moods and desires of their animal partners and reward them with food, rest or other treats.[24] Renfro pampered his performing dogs, building them 'their own bathhouse, with three large tubs', exercising them every day in a 'peach orchard or on a special running machine' and feeding them on 'a ground mash consisting of 70 per cent meat, 30 per cent vegetables'.[25] The Mughal emperor Akbar (1542–1605) fed his best-performing hunting chee-tahs around 2.5kg of mutton per day (a generous amount), honoured them with decorated collars and leashes and gave each animal his own personal staff of 3–4 keepers to attend to its daily needs.[26] More dramatically, labour-ing animals can sometimes exercise a degree of agency in their working lives by actively withholding their labour, forcing their human co-workers to take their needs and preferences into account. Turnspit dogs (which usually worked as a pair), would often decline to operate the wheel if made to do so on their day off.[27] Elephants in a timber yard at Rangoon insisted on having 'a regular working day of fixed hours and a regular scale of payment in the shape of food and such luxuries as sugar cane', going 'on strike' if their requirements were not met.[28] Cows in early modern England might refuse to be milked if approached from the wrong side or treated roughly.[29] Animal labour can therefore, under certain conditions, be beneficial – or at least tolerable – to non-human workers, and its termination, rather than liberat-ing animals from drudgery, may result in their obsolescence or slaughter.

Does animal labour facilitate or diminish the exploitation of humans? Here, once again, the jury is out. In some cases, of course, animal labour has undoubtedly assisted humans, making their work less arduous or enabling them to accomplish things they would have been unable to do without ani-mal strength or agility. Ploughing fields and transporting goods, for instance, became easier in colonial New Spain and Peru following the introduction of horses and oxen from Europe (though these gains were largely negated by the recruitment of indigenous people to work in plantation agriculture and mining). Camel transportation replaced human porters in the medieval Middle East, while guide dogs have enabled many visually impaired people to lead independent lives. Animal labourers have also done jobs that might otherwise have exposed humans to injury or death, either because their physical characteristics render them less likely to be killed, or because their lives are deemed more expendable; in 2020 an African giant rat named Magawa received a gold medal from a British animal charity after sniffing out 71 land mines in Cambodia, his small body size making him less liable to trigger an explosion than would a human being.[30] In all of these cases, ani-mal labour has benefited human workers, making their lives easier.

Ranged against these positive outcomes of animal labour, however, we have plentiful examples of animal work actively facilitating human exploita-tion, or of animals and workers being pitted against one another by their employers. In the colonial Caribbean, for instance, oxen, horses and African slaves laboured together on sugar plantations, with the former unwittingly encouraging the abuse of the latter by making their enslavement profitable; tellingly, both were referred to as 'stock' – an indication of the perceived interchangeability of human and animal labour in the eyes of plantation owners. In Cuba, Saint Domingue, Jamaica and the southern states of the USA, meanwhile, dogs contributed directly to human oppression by tracking down indigenous people and escaped slaves, and even, during the Haitian Revolution, eating rebel maroons, thus acting as guardians of the slave system.[31] The French General Donantien de Rochambeau, for example, imported two hundred bloodhounds from Cuba to use against Toussaint L'Ouverture and his troops, allegedly feeding the animals with the flesh of captured rebels.[32] From the nineteenth century, moreover, both animals and humans suffered increasingly from a productivity-driven capitalist system that encouraged exploitation and over-work, with impoverished humans often taking out their frustrations on the non-human species that worked alongside them. William Peacock, the 'ragged-looking fellow' fined for hitting a donkey with a stick at Fairlop Fair, claimed that he only beat the animal because 'the young gentleman who was riding wished it to go faster' – a sign of the class dynamics at play in human–animal working relationships. Thomas Horsoy, the omnibus driver prosecuted in 1841 for working an injured horse, begged the magistrate 'to get him another situation' following his conviction, 'for if he was not to drive horses provided by his employers he might as well give up the occupation of driver'.[33] Animals and humans could therefore be allies, working together to achieve a common goal (though in most cases one set by humans), but they were also often victims of wider, capitalist systems of oppression that set them against one another.

Co-workers or Slaves?

To explore these issues in greater depth, we will look at three case studies, each of which illustrates a different aspect of animal labour. The first, a global history of the donkey, shows how one unassuming animal changed the world – while being largely ignored. The second, the use of dog carts in nineteenth-century Britain, highlights the questionable morality of making a popular companion species perform tiring and physically injurious labour, and the (sometimes unexpected) implications of outlawing controversial forms of non-human work. The third, the recruitment of animal astronauts in the mid-twentieth century, illuminates the complex ethics of using non-human actors to undertake work deemed too dangerous for humans.

Donkey Work

Donkeys have played a key role in farming and trade since their domestication from the African wild ass (*Equs africanus*) around 7,000 years ago.

Hardy, versatile, and capable of travelling long distances without dehydrating, donkeys opened up desert regions and connected cities and markets. They were also the first animals to be ridden. Historian Brian Fagan describes them as 'catalysts for change' and an 'instrument of globalisation'.[34]

Donkeys first made their mark in the ancient world, where they worked in agriculture, carried goods and people and played an important role in overland trade. In ancient Egypt donkeys transported gold and tropical products from Nubia, turquoise and copper from the Sinai and ivory from West Africa. In the Middle East donkey trains carried textiles and metals between Assyria and Anatolia. In Classical Rome donkeys were used to crush olives, press grapes, grind grains and transport goods across the empire.

From the Mediterranean, donkeys spread to northern Europe, carrying pepper, cloves, ginger and salt across the Alps. In the fifteenth and sixteenth centuries they arrived in the Americas, where they were employed to transport silver from the mines at Potosí and Zacatecas to the coast. By the nineteenth century they were traded widely across the Indian Ocean World, working on clove plantations in Zanzibar, operating pumps in South African diamond mines and lugging slabs of salt from the Danakil Depression to the Ethiopian Plateau.[35] Donkeys also moved goods over shorter distances, serving as the main form of transportation for farmers and artisans. Visiting Persia in the 1890s, American Samuel Graham Wilson noted the prevalence of donkeys on the streets of Tabriz, where they carried 'provision baskets filled with grapes … garden truck or firewood' and hauled 'all sorts of building materials, the bricks falling off, the poles dragging behind, and ready to hit an unwary pedestrian in the shins'.[36]

While donkeys thus facilitated trade and boosted agricultural production on a global scale, the precise nature of their relationship with humans has been debated. Did their employment exacerbate or diminish human exploitation? Were they respected partners or exploited slaves?

If we start with the question of human exploitation, the picture appears mixed. On the one hand, donkeys, by facilitating the long-distance transportation of prestige goods, enhanced the power of elites, helping to embed social hierarchies in the ancient world. Along with horses and mules, donkeys also played a crucial role in the conquest and retention of empires, facilitating the transmission of information across vast distances and keeping imperial armies supplied with rations and equipment. As Peter Mitchell observes, donkeys and mules 'exported gold from Brazil's Ouro Preto mines to the Atlantic coast, carried Bolivian silver across the Isthmus to Panama, connected Mexico City to the ports of Veracruz and Acapulco, and helped Euro-American settlers to colonize what is now the United States'.[37]

On the other hand, while donkeys often served as unwitting collaborators of empire, there were also instances in which their introduction permitted indigenous people to maintain their independence or improve their agricultural yields. In Southern Africa the adoption of donkeys by Bushman peoples of Botswana helped them to extend the range of their hunting and foraging expeditions. In Jujuy, northern Argentina, Native Americans played an important part in supplying mules for the transportation of goods across the Andes. They earned enough money from this work to pay their tribute to the Spanish authorities in cash, relieving them of the labour obligations inflicted upon other indigenous communities.[38] Across the globe, donkeys have also reduced workloads for their (often poor and/or female) owners, helping them to plough fields or take produce to markets; one Spanish donkey carried 'canisters of milk' into Madrid when his owner fell ill, 'r[inging] all the house bells where he had to call with his teeth to make it known that he was there'![39] Donkeys could thus perpetuate the power of elites, but they could also help individuals and groups to alleviate poverty or resist colonialism.

As for donkeys themselves, they have also had varied experiences - though the scales tip in the direction of exploitation. On the positive side of the ledger, donkeys appear to have been valued highly in the ancient world and were sometimes buried alongside their owners. One pharaoh was buried alongside ten donkeys at Abydos around 3000BC – an act which suggests a degree of respect for the animals (though they may well have been sacrificed for the purpose).[40] A noblewoman from Xi'an in Tang dynasty China was likewise interred alongside three donkeys following her death in 878AD (archaeologists believe she may have used the animals for playing polo).[41]

On the negative side, many donkeys were treated poorly by humans, suffering, overwork, abandonment and death. Clay tablets from the city of Kültepe in central Turkey suggest that donkey caravans travelling between Assur and Kanesh around 1900BC sometimes suffered mortality rates of up to 50%.[42] Visiting Egypt in 1901, Briton T.M. Greg complained that the donkeys used by tourists were forced to wear painful bits in their mouths and 'frequently beaten about the head and dragged by the bridle to make them change their place or run after a prospective employer'.[43] When motor trucks supplanted donkeys in South Africa's diamond fields in the 1920s, meanwhile, many of the latter were 'ruthlessly abandoned by their owners', who, rather than incur the expense of watering and feeding them … have turned them out on the arid veldt to shift for themselves'.[44] Though a crucial cog in multiple industries, therefore, donkeys have often been viewed as replaceable and expendable, cared for by some, but underappreciated by most.

The story of the donkey illustrates the degree to which animal labour has shaped civilisations. Less celebrated than horses, donkeys have received little attention from historians, and have often been omitted from the historical record.[45] Despite their humble status, however, these anonymous equines have left a substantial hoofprint on many human societies, forging regional and transcontinental connections and facilitating global and local trade.

The Dog Cart Nuisance

Dog-drawn carts were a common sight in early modern Europe. A cheaper alternative to horse-drawn carriages, dog-drawn vehicles were used by hawkers to transport their wares around town, serving, most frequently, as conveyors of milk and fresh bread. They were especially prevalent in Belgium and the Netherlands, appearing in paintings and photographs from the sixteenth to the early twentieth centuries (Figure 2.1).

In the 1830s dog carts started to attract heavy criticism from a growing band of British humanitarians, who viewed them as both dangerous and cruel. Critics argued that dog carts posed a threat to other road users, due to reckless driving, and that overworked canines were more susceptible to rabies – a disease much feared at the time. Opponents of the so-called 'dog cart nuisance' also emphasised the suffering experienced by the working dogs, whose anatomy was unsuited to long journeys on paved roads. Portsmouth councillor Major Travers urged 'any gentleman' to 'examine, as I have done, the feet of a dog after having been driven a considerable distance on a hard and gravelly road, and he will find them lacerated and swollen, and in fact covered in raw sores'.[46] The Bishop of Oxford contended that the 'foot of a dog was of a soft and spongy nature, most admirable for the purposes that nature intended it, but certainly unsuitable to draw heavy loads over hard and flinty roads'.[47] Moved by such arguments, Parliament

Figure 2.1 'A Dog in the Wrong Place', *The Animals' Friend*, 1913, p. 119.

Source: photograph by author.

prohibited the use of dog carts in 1854, outlawing a practice seen as brutal and outdated.[48]

The abolition of dog carts highlights some of the complexities associated with animal labour and the ethical dilemmas posed by its removal. Viewed on its own terms, the measure was a well-meaning one, designed to protect vulnerable animals from abuse. Many canine hauliers endured poor working conditions, succumbing to accidents, exhaustion and underfeeding. Some were mistreated by their owners, suffering the effects of brutal over-driving. In 1846, for instance, magistrates convicted Thomas Poding, a razor-grinder from Bromyard, for having 'furiously and cruelly driven a dog tackled to a machine for grinding razors, &c.', 'riding on the machine without having any reins upon his canine slave, and going at the speed of eight or nine miles per hour along the Hereford Turnpike-road'.[49] In 1854 magistrates in Gosport sentenced John White, a rag and bone collector, to 'one month in the House of Correction, with hard labour' for beating two dogs 'unmercifully, at the same time jumping up and down in the cart and swaying the whole weight of his body to and fro, so that the dog attached to the shaft ... was nearly crushed to the ground'.[50] Banning the use of dog-drawn vehicles was designed to end this kind of cruelty, and to remove the pitiful sight of panting, bleeding canines from British streets.

Taking a longer view, the abolition of dog carts proved to be a contentious measure which ended up having some unforeseen negative consequences for both humans and dogs. From the human perspective, the new legislation criminalised what was, for many, an important source of income, making it harder for some of the poorest in society to earn a living. In 1846, for example, magistrates fined 'John Ward, an old man, seventy years of age, who hawks china, glass and fruit about Surrey' for using two dogs to pull a 'heavily laden' barrow down the Clapham Road. Ward, who was 'three parts blind and almost a cripple', could not pay the fine (1 shilling), so was sentenced to seven days in prison, losing both his livelihood and his liberty.[51] From a canine perspective, the abolition of dog carts seemed like a blessing, but led, in many cases, to the culling or abandonment of working dogs, who were discarded by their former owners when they ceased to have value as labourers. Instead of being granted a comfortable retirement, as intended, they were simply killed or left to starve on Britain's streets. Other species, moreover, suffered indirectly as a result of the abolition of dog carts, filling roles previously performed by canines. In 1855, for instance, police in Stamford arrested 'a fellow named W[illiam] Cooper, a higgler from Loughborough', for 'using worn-out and diseased horses' to pull a heavy waggon - a practice that had reportedly 'become very prevalent since the abolition of the dog cart nuisance'.[52]

The dog cart case thus serves as a warning of what can happen when the working partnership between humans and animals is dissolved. The middle-class humanitarians who campaigned against the use of dog carts acted in what they believed to be the animals' best interests. Their advocacy, however, had unintended results, leading to both human and animal suffering. This pattern would be repeated across the globe, as changing ethics and new technology put growing numbers of animals out of work. In 1997, after the Burmese supreme courts imposed a ban on timber exports, 'about 100 elephants used for timber transportation' were rendered 'jobless', experiencing not liberation, but hunger and abandonment.[53]

Animal Astronauts

Dog carts were an old technology, rendered obsolete by economic change and shifting attitudes towards animal labour. By contrast, the mid-twentieth century witnessed the use of animals in the development of a very new technology – the space rocket – as the USSR and the USA competed to launch the first humans into orbit. The Russians primarily used dogs in their space programme, while the Americans enlisted monkeys and chimpanzees on account of their higher cognitive capacities and closer physiological resemblance to humans. In both cases the use of animals in space research posed challenging questions about the nature and ethics of animal labour and the status of canine and primate astronauts. Were they active labourers or passive experimental subjects?

Animals first started to be used in space exploration in the late 1950s and continued to play a role until the 1980s. In 1956 the Russians sent two dogs 'dressed in space suits' to a height of 68 miles and parachuted them safely back to Earth.[54] In 1957 a third dog named Laika became the first living thing to orbit Earth, allowing scientists to collect data on 'the response of the animal's heart and lungs to weightless existence in space'.[55] In 1960 two further dogs, Strelka and Belka, circled Earth 17 times to test the physiological effects of prolonged weightlessness.[56]

On the other side of the Atlantic the Americans launched several monkeys into space in the late 1950s, among them rhesus monkey Able and squirrel monkey Baker, 'fired to an altitude of 300 miles' to test the effects of acceleration, deceleration and weightlessness on the body.[57] NASA later expanded the programme to include chimpanzees, sending two of the animals into space in 1961. The first, a three-year-old chimpanzee named Ham, survived a flight to an altitude of 155 miles in a Mercury capsule in February that year, 'working simple controls while strapped onto a 'contoured couch'.[58] The second, a young male named Enos, orbited Earth twice in an Atlas rocket and was safely retrieved from the sea off Bermuda. The latter experiment was designed to determine 'whether an American astronaut could carry out manual tasks while undergoing prolonged weightlessness'. It was followed a few months later by astronaut John Glenn's first trip into space.[59]

The decision to use dogs and primates as astronauts raised questions about the nature of their labour and the morality of using sentient (in this case, highly intelligent) beings to perform dangerous tasks. Did the animals qualify as workers? Did they suffer during their assignments? Were they rewarded for their efforts? The evidence is mixed and highlights the complexities of defining and recognising animal labour.

If we consider first whether the activities undertaken by canine and primate astronauts constituted labour, the answer would appear to be yes. Soviet and American scientists selected dogs and primates for the early space missions because they were intelligent enough to operate controls and physiologically close enough to humans to replicate effects of space flight on the human body. This meant that they were trained for their work and, to some extent, respected for what they did. Enos, for instance, learned 'to punch levers upon a signal from a panel equipped with lights and symbols to get ... rewards or prevent a mild reminder electrical shock'.[60] A chimpanzee named Bobby Joe at a training facility in Alamogordo exhibited exceptional levels of cognition, 'using two levers to keep a small cross inside a lighted circle on a lighted panel board, one possible spacecraft control system'.[61] This behaviour required a degree of cooperation between humans and primates and was explicitly seen by the former as work; Enos was described in one report as 'a real beaver for work', continuing to operate his levers throughout his flight.[62] Good performance was also in most cases directly rewarded with a treat, giving the primates' labour a transactional quality. Bobby Joe received a 'banana-flavoured food pellet' every time he kept the cross inside a circle for two 10-second periods', earning 225 pellets during one '5-hour shift' (the term 'shift' also denotes work).[63]

While the contributions made to the space programme by canine and primate astronauts can therefore be classified as work, the conditions under which that work occurred raised serious ethical issues, placing the animals in significant danger and threatening their physical and mental well-being. First, of course, and most well-publicised, some non-human space-travellers perished in accidents, dying in explosions or drowning in the sea when rescuers failed to locate their capsules. Squirrel monkey Gordo, for example, was lost at sea following 'a mishap to the recovery apparatus of the missile's nose-cone', while Able died under anaesthetic during an operation to remove an electrode from under her skin.[64] The Russian dog Laika perished in space from overheating; fellow dog Lisichka died when her rocket exploded on the launchpad.[65]

Second, and less well understood at the time, some space veterans suffered significant physical and psychological trauma as a result of their experiences, in part because the demands put on them in flight exceeded what they had been prepared for by their training. Ham, for instance, exhibited 'a little fatigue, a little wobbling and trembling of the legs and a slight abrasion on his nose' when he was retrieved from his travelling capsule, and subsequently 'balked, screeched and hugged his handler's neck when two veterinarians tried to insert him into a metal couch similar to the one he rode on his space flight'.[66] A 'fault on the stabilizing system' on Enos's flight, meanwhile, meant that 'he was subjected to a continuous rocking sensation', leaving him with 'hypertension ... caused by frustration and confusion'.[67] Such reactions indicate that dogs and monkeys were being pushed to the limits of their capabilities and forced, without their consent, to undertake risks that would not have been taken with humans.

The experience of primate and canine astronauts illustrates some of the tensions and complexities inherent in animal labour, and the ethics of employing non-humans to perform dangerous tasks. The dogs, monkeys and chimpanzees trained by NASA and the Soviet Space Programme were taught to respond proactively to specific stimuli, and were explicitly perceived to be working. The animals' value to humans grew directly in proportion to the level of training they had received (chimpanzees were said to cost $1,000 each, and to become 'much more valuable' after training), and those individuals who achieved space firsts were (in some cases) given a pampered retirement and honoured posthumously.[68] Ham, for instance, became a national hero and was buried in the International Space Hall of Fame in New Mexico after his death in 1983; scientist Oleg Gazenko adopted space-dog Zhulka, the survivor of an unsuccessful launch in 1960.[69] Ultimately, however, these non-human workers were considered expendable – or at least more so than their human counterparts – and the purpose of their work was to act as proxies for human astronauts. They did not have a choice about their inclusion in the space programme, and many of the less high-profile animals were subsequently transferred to zoos or medical facilities, losing the privileges of a working animal. The line between labourer and experimental subject was therefore a fine one, and the dogs' and primates' status as workers impermanent.

Animals at War

So far, we have focused on the role of working animals during peacetime. There was, however, another, darker side to animal labour that merits more in-depth discussion: the use of animals in war. Long recognised by military historians as key actors on the battlefield, animals – particularly horses – have on many occasions determined the outcomes of conflicts and facilitated the conquest of states. They have also, of course, been victims of warfare, perishing in large numbers from wounds sustained in battle and (more commonly) from malnutrition and disease. The role of animal labour in war, therefore, holds a particularly important place in world history and poses especially challenging questions regarding the ethics of recruiting non-humans for dangerous work. What roles have animals performed in war? How has their contribution been perceived and valued? To what extent has the use of animals for military purposes shaped world history?

Animal labour in war can broadly be divided into three categories: combat, logistical support and ceremonial use. First, at the more visible end of the spectrum, animals have participated in active combat. Horses have for centuries carried human soldiers into battle, initially in specially designed war chariots, later as mounted cavalry, giving troops a major speed and height advantage over unmounted infantry. Camels and elephants have also played an active part in military campaigns, sometimes to devastating effect. Camels, for instance, participated in Xerxes' Second Persian invasion of Greece (480–479BC), and, more recently, served in the British-officered Egyptian camel corps at the battle of Omdurman (1898). Elephants were prominent in Indian, East Asian and North African warfare until the early eighteenth century and even, on occasion, operated in Europe. The Carthaginian leader, Hannibal, famously brought 37 elephants from Spain across the Rhone and over the Alps into northern Italy to fight against the Romans in the Second Punic War (218–201BC).[70] Though neutralised from the mid-nineteenth century by increasingly deadly artillery, animal combatants have had a significant impact on warfare and have sometimes changed the outcome of battles. A well-executed cavalry manoeuvre by General José Antonio Páez routed royalist forces at the battle of Las Queseras del Medio (1819), consolidating Venezuela's independence from Spain.

Less well-documented in the chronicles of war, but equally important, animals have played a significant part in the logistics of warfare, carrying rations, hauling artillery and ferrying messages between troops. In 1532 the Inca general Quizquiz abandoned 15,000 llamas in the eastern mountains after a battle in the first years of the Spanish conquest – an indication of the large numbers employed in supporting the Inca army.[71] In 1944, during the Japanese occupation of Burma, Lieutenant-Colonel James Howard Williams assembled a company of 'more than 100 elephants' to build bridges across 'the shallower rivers and streams' of the Kabaw Valley, facilitating the movement of British troops in the region.[72] During World War I, the British army trained dogs – mostly Airedales – to carry a form of field telephone, allowing officers to relay commands over a distance (Figure 2.2). Though less glamorous than mounted cavalry, pack animals, carrier pigeons and bridge-building elephants were all vital to the success of military campaigns and, in many cases, made military victories possible. Despite the growing use of trains, tanks and motorised vehicles in modern warfare, pack animals remained central to army supply systems until well into the twentieth century, lugging food, ammunition and medical equipment across challenging terrain. When Adolf Hitler's forces invaded the Soviet Union in 1941, they employed 750,000 horses to pull the artillery, supply wagons, field kitchens and ambulances needed by the Wehrmacht.[73]

Figure 2.2 'For Telephone Service', from 'Animals in Warfare', *The Animal World*,
        November 1914, p. 198.

Source: photograph by author.

Animals have also played a more ephemeral role in war as army and navy
mascots, entertaining soldiers and sailors in barracks and aboard ship and
providing a welcome distraction from the tension and boredom of campaigns. When villagers from Mérida in Venezuela presented General Simón
Bolívar with a young Mucuchí dog in 1813 he kept the animal by his side as
a companion until the canine's death in 1821 at the Battle of Carabobo.[74]
When the city authorities of Albany, Australia, presented the officers of the
US battleship *Wisconsin* with a kangaroo named Murphy, the crew adopted
the marsupial as a mascot, training him to box with his fellow sailors.[75]
Though often dismissed as frivolous luxuries, military mascots sometimes
made a significant contribution to the war effort, boosting morale in challenging conditions and, in many cases, forging a close bond with the men
who lived alongside them. Some mascots, indeed, faced real danger while
engaged at the front, suffering injury or death at the hands of the enemy.
Simon, the ship's cat aboard HMS *Amethyst*, sustained serious injuries in
1949, when the vessel came under fire on the Yangtse River during the

Chinese Civil War, and was subsequently awarded the Dickin Medal for killing vermin while the ship lay under siege (the only cat ever to receive the
honour).[76] Animals thus performed both physical and emotional labour during war, providing human soldiers with mounts, intelligence and munitions
and offering some light relief from the stresses of warfare.

How far has the use of animal labour for the purposes of warfare impacted
on human history? What was the experience of war like for animals themselves? To what extent were animal co-belligerents recognised for their contribution to human conflicts? These questions, applicable to animal labour
more widely, have particular resonance in the field of military history, where
the stakes were high for both animals and humans.

If we consider, first, the importance of animals in war in a world history
context, it becomes clear that their labour has changed not just the outcomes
of battles, but of entire campaigns. This has had major consequences for the
survival of states and the expansion and contraction of empires. In sub-
Saharan Africa, for instance, the introduction of horses to Senegambia in
the fifteenth century by Islamic traders transformed the nature of warfare in
the region, giving rise to politically centralised states such as Mali.[77] In medieval India the Muslim invaders who established the Delhi Sultanate in the
thirteenth century owed their battlefield successes over Hindu armies to
their efficient supply of war horses and elephants, which enabled them to
defeat armies far larger than their own.[78] In thirteenth-century Central Asia
the infamous Mongol warlord Genghis Khan owed many of his conquests
to mastery of the horse, which allowed his mounted warriors to overrun
large swathes of the continent and establish one of the largest empires in
history. The physical capabilities of the horse also set the territorial limits on
Khan's Eurasian empire, which might have extended further into Europe
had sufficient fodder been available for his equine co-combatants.[79]

For a more modern example of how animals have influenced the outcome
of wars, we might take the case of the American Civil War, where Union
victory can be attributed, in part, to the supply of horses. As Gervase
Phillips has shown, the Confederacy enjoyed an early advantage on account
of their soldiers' more extensive acquaintance with equines. Many south-
erners brought their own horses to the front lines, which meant they were
more familiar with the animals' movements and better able to control them.
Most Union soldiers, conversely, had little experience of riding and struggled to manoeuvre their horses in battle. As the war wore on, however,
Federal control of key horse-breeding areas in Kentucky, Missouri and
Tennessee cut off the South from their supply of mounts and made it impos-
sible for them to replace horses lost to injury or disease. Together with a
severe shortage of horseshoes, this hampered the mobility of Confederate
armies, forcing General Robert E. Lee to fight a static campaign around
Petersburg from 1864. Union General Philip Sheridan, by contrast, was able

to traverse the Shenandoah Valley on horseback with his troops, defeating Confederate armies and destroying much of the South's food supply. Access to horses was thus crucial to Union success in the American Civil War and had a direct influence on military tactics. Fatal cavalry charges such as George Pickett's doomed assault on Union lines at the Battle of Gettysburg (1863), meanwhile, highlighted the growing redundancy of horses in combat, and the ineffectiveness of mounted riders against increasingly deadly artillery fire.[80]

The impact of warfare on animals themselves has been little examined until recently, but is starting to be reassessed by animal studies scholars. Primarily, of course, the animal experience of warfare was often one of extreme suffering – a fact attested by the high animal casualties recorded on military operations. In 1817, for example, when Argentine General José de San Martín crossed the Andes to liberate Chile from Spanish rule, 1,100 horses and 4,981 mules perished in the journey, some falling to their deaths and others succumbing to the cold.[81] During the Second Afghan War (1878–1880) 60,000 camel transport animals were lost by the British army, succumbing to a combination of overloading, under-feeding, disease and poor veterinary care.[82] In the Boer War – widely regarded as one of the worst cases of horse attrition in any conflict – 326,073 horses and 51,399 mules died between October 1899 and May 1902 on the British side alone, representing death rates of 66.88% and 35.37% of the total head count, respectively. The causes of these shocking death tallies included injury from enemy fire, overwork, malnutrition and substandard care. Disease was also a major killer of military animals, its spread exacerbated by weakened immune systems and lack of proper rest. Sandra Swart attributes the shocking equine casualties in the Boer War to the importation of large numbers of horses from Europe, Australia and the Americas, many of which arrived dehydrated, malnourished and with their immune systems severely compromised, making them susceptible to mange, glanders and pneumonia.[83]

While warfare was clearly traumatic for most animal participants, it could, nonetheless, foster more positive human–animal relations and engender respect for animal combatants – in the same manner as other close human–animal partnerships. Forced to work closely with non-human animals, and, in many cases, dependent on the latter for their survival, soldiers and sailors often formed intimate bonds with creatures they trained or rode and expressed admiration for their bravery and skill. Some turned to their non-human companions for friendship or solace. Some mourned the passing of equine or canine comrades as valued partners Others celebrated the intelligence and initiative of specific animal co-belligerents, emphasising their valour or fidelity. The sailors on board the USS *Wisconsin*, for instance, showed great affection for their marsupial mascot, Murphy, declaring that 'a smarter kangaroo never hopped, skipped or jumped'.[84] A corporal in the

Royal Garrison Artillery during World War I expressed similar fondness for his pet dog, who served alongside him in the trenches:

> It is a fox terrier, and I found him in a dug-out on the old Somme battlefield, half-starved, and since then he has shared my meals and also my blankets at night. He is such a faithful companion and I cannot think of leaving him behind when I return to Blighty.[85]

Such touching words demonstrate a degree of affinity with animal comrades and an awareness that their actions could improve morale among the troops. War was therefore an unnatural and stressful experience for animals, but it also offered opportunities for recognition, commemoration and camaraderie.

Horses in the Spanish Conquest of Mexico

To explore the contribution to animals to human war in more detail we will look an event in which horses and (to a lesser extent) dogs, played a pivotal role: the Spanish conquest of Mexico (1521). One of the most dramatic conquests in world history, the defeat of the Aztecs by a band of Iberian mercenaries changed the history of the Americas and initiated a period of colonial rule that would last for three hundred years. Though by no means the only factor in the conquistadors' triumph, horses played an important part in ousting the Aztecs and exerted an influence out of proportion to their numbers. How, then, did horses assist in the conquest of the Aztecs? How highly were they valued by the Spanish and how were they perceived by indigenous Mesoamericans? To what extent did their presence influence the outcome of the contest?

When Cortés disembarked in Veracruz in 1519 he had only sixteen horses in his landing party. Despite being few in number, however, the conquistador's equine auxiliaries gave his troops a significant advantage over their Mexican adversaries, allowing them to slash at the enemies from a height with deadly steel swords (also absent from the Aztec armoury) and to launch devastating cavalry charges. Often heavily outnumbered by their opponents, the Spaniards appear to have been saved from defeat on several occasions by well-timed intervention from their cavalry, which, in many cases tipped the balance in their favour. Cortés, for instance, claimed that the Aztecs in Tenochtitlán 'were frightened by the horses and began to flee' when they first saw them.[86] Bernal Díaz, an ordinary soldier in Cortés's band, recounted how another indigenous group (and later important Spanish ally) the Tlaxcallans, 'fought all the time like brave warriors, until the horsemen came up', upon which their resistance quickly collapsed. According to Díaz, this was because they 'thought that the horse and its rider was all one animal, for they had never seen horses up to this time'.[87]

As Díaz's comments indicate, it was not merely the military capability of the horses that made them such a powerful weapon against the Indians, but also the psychological impact of seeing these imposing beasts for the first time. Mesoamerica possessed no large herbivores of equivalent stature to the horse (the largest mammal in Central America is the tapir) and the Aztecs and their Native American counterparts had no experience of riding animals into battle. The sight of a mounted warrior galloping towards them at full speed must therefore have been terrifying for indigenous Mexicans, all the more so if they believed, as Díaz alleged, that the horse and rider comprised a single living creature. The conquistadors, furthermore, quickly grasped the potential of horses to induce terror among Mesoamericans, making deliberate efforts to maintain and enhance their mystique. Diaz, for instance, chronicled how Cortés played a trick on several indigenous leaders by bringing a stallion before them and allowing him to scent a mare who was tethered nearby, causing the animal to 'paw the ground and neigh and become wild with excitement'.[88] Later in the campaign, when messengers arrived from the Aztec Emperor Moctezuma II, Cortés ordered his lieutenant Pedro de Alvarado and the other horsemen to gallop along the beach, attaching 'little bells to the horses' breastplates' to increase the drama of the spectacle.[89] All of this was designed to reinforce the horses' quasi-supernatural status.

A closer analysis of the writings of the conquistadors underlines the extent to which they recognised the contribution of horses to their campaign and the degree of esteem in which they held the animals. Letters and memoirs written by the soldiers repeatedly mention horses, which, rather than being described en masse, are often individuated, suggesting that they possessed specific characteristics that rendered them suitable for warfare. Díaz, for example, furnished a detailed account of the sixteen horses purchased on Cuba prior to the Mexican venture, recording their sex, colour and character: Pedro de Alvarado's horse was 'a very good sorrel mare, good both for sport and as a charger'; Diego de Ordaz had 'a gray mare, barren, tolerably good, but not fast'; Gonzalo Domínguez acquired 'a good chestnut horse, a grand galloper'.[90] When the conquistadors' horses were injured or killed, moreover, there is evidence that the Spaniards attempted to medicate or conceal the animals, preventing their enemies from realising that they were mortal. Cortés openly grieved the loss of a mare during the siege of Tenochtitlán, but was 'pleased that she had not perished at the hands of the enemy, as we thought would happen, for their joy at having captured her would have exceeded the grief caused by the death of their companions'.[91] Díaz, meanwhile, recounted how, following the clash with the Tlaxcallans, 'we doctored the horses by searing their wounds with the fat from the body of a dead Indian', preserving the precious animals to fight another day.[92] These were clearly the actions of men who appreciated the military value of their steeds and considered them vital to the campaign. As Cortés expressed it: 'after God we had no help save from the horses'.[93]

As for the Aztecs and other Mesoamerican peoples, it is harder to know how they perceived their enemies' equine allies, though the limited source material available corroborates Spanish claims as to their significance. On the one hand, Mesoamericans were evidently intrigued by the conquistadors' horses (which the Aztecs referred to as *mamazah*, or 'deer', in their language) sending artists to study the novel animals and according them a prominent place in subsequent chronicles of the conquest.[94] According to Díaz, Moctezuma sent 'clever painters' to Cortés's camp in order to 'make pictures true to nature' of the conquistador, his soldiers, and, significantly, his 'ships, sails and horses' – all crucial Spanish technologies of war.[95] The Lienzo de Tlaxcala (a cotton cloth showing scenes from the conquest) likewise depicts mounted and tethered horses, paying close attention to their muzzles, saddles and shoes and including visible hoof-prints on the pathways that snake across the image (human footprints were used on Pre-Columbian maps to indicate direction of travel).[96] On the other hand, it seems that indigenous people quickly came to understand the conquistadors' mounts as a threat to their existence and, where possible, took measures to neutralise or eliminate them. Cortés reported that the Aztecs filled the main square in Tenochtitlán 'with boulders to stop the horses crossing it'.[97] Díaz claimed that the Indians in Cholula (Aztec allies) dug holes in the streets and filled them 'with sharp pointed stakes to kill the horses when they galloped' – clear evidence that the latter no longer viewed the horses as immortal and had adapted their tactics accordingly.[98] The Aztecs also sacrificed captured horses alongside human warriors, flaying their skins and displaying their severed heads on skull racks (*tzompantli*) alongside those of their human riders; the Florentine Codex shows four horses' heads arrayed on a *tzompantli*, below eight bearded heads of human conquistadors.[99] This suggests that the Aztecs viewed horses as equal in stature to their human enemies and similarly worthy for sacrifice to the gods.[100] It has, indeed, been proposed that the Aztecs admired the horses more than the Spaniards because the former charged bravely into battle, and, unlike the latter, did not wear armour.

Looking at the conflict overall, therefore, it is clear that Spanish access to horses made a material difference to the conquest of Mexico and played a significant role in Cortes's success. Horses, of course, were not the only factor in the defeat of the Aztecs. Superior weaponry, the collusion of indigenous allies and the devastating effect of Old World diseases on New World peoples were equally critical to Spanish victory, working together to undermine Aztec resistance and morale. Cultural differences in the waging of war also had an impact on early skirmishes. Equine allies were, nonetheless, a valuable and very visible tool in the conquistadors' armoury, and both Spaniards and Amerindians appreciated their military significance.

Ultimately, indeed, Aztec efforts to counter Spanish horses by blocking streets and digging pits forced Cortés to level Tenochtitlán, with horrific consequences for its indigenous defenders.

In the longer term, the relationship between horses and Spanish colonialism in the New World proved more ambiguous. In Peru, the site of a further Spanish conquest in 1532, horses again played a key role in subduing a major Pre-Columbian civilisation – in this case the Incas – exerting a similarly devastating impact on unmounted troops; a drawing by the indigenous chronicler Felipe Guaman Poma de Ayala depicts an armoured conquistador lancing the Inca general Quizo Yupanqui Inca from the back of a grimacing horse (Figure 2.3). In more distant parts of the Americas, however, feral horses introduced by the Spanish reached indigenous peoples before any Europeans did and were enlisted by Amerindians to defend themselves against Spanish marauders - in some cases with considerable success. The Mapuche in southern Chile managed to retain their independence into the mid-nineteenth century with the help of horses, fighting off Spanish troops and, later, conducting raids against Spanish settlements. The Chichimeca in northern Mexico also became effective horsemen, fighting a forty-year contest with the Spanish and stealing European livestock.[101] Initially a decisive weapon in Spanish colonialism, therefore, the horse gradually lost some of its power as Native Americans incorporated it into their arsenals, becoming central to indigenous military operations as well as to European ones.

Conclusion

Animal labour has played a critical role in the development of human societies. Oxen have ploughed fields. Camels, mules, llamas and donkeys have facilitated transcontinental trade. Horses have carried human riders in peace and war, while dogs have herded sheep, pulled carts, carried messages and turned meat on a spit. A range of other species have performed more bespoke jobs, from locating truffles to assisting the disabled; one baboon in 1890s South Africa operated the points at a railway station in the colony, '[t]he man in charge of the latter having in a railway accident, lost one arm and part of the remaining hand'.[102]

Given their sustained interactions with human handlers, working animals offer particularly interesting insights into the human–animal relationship, as well as raising broader questions about the nature of labour. To what extent can animal labourers shape their own working conditions? Are working animals valued partners or exploited slaves? Can working animal go on strike, if assigned tasks that cause discomfort or fear? Can they retire? Studies of animal labour in a historical context reveal a nuanced picture of exploitation, abuse, collaboration and appreciation, in which human and animal labourers negotiate, test, and sometimes resist the conditions of their employment.

Figure 2.3 Drawing 157. Captain Luis de Ávalos de Ayala kills Quizo Yupanqui Inka in the conquest of Lima. Guaman Poma, *Nueva corónica y buen gobierno* (*c*.1615), p. 392.

Source: Royal Danish Library, GKS 2232.

Notes

1 Alan Mikhail, *The Animal in Ottoman Egypt* (Oxford: Oxford University Press, 2014), p. 49.
2 Verene A. Shepherd, 'Livestock and Sugar: Aspects of Jamaica's Agricultural Development from the Late Seventeenth to the Early Nineteenth Century', *The Historical Journal* 34:3 (1991), pp. 627–643.
3 Jonathan Saha, *Colonizing Animals: Interspecies Empire in Myanmar* (Cambridge: Cambridge University Press, 2022), pp. 28–50.
4 Antonio León Pinelo, *El Paraíso en el Nuevo Mundo* (Lima, 1943) vol. II, p. 53.
5 'The Alpacas', *The Sydney Morning Herald*, 21 August 1860.
6 'Addis Ababa and its Hyenas have a Long and Peaceful History', *The Guardian*, 5 March 2014.
7 'Novel Mowing Machines for Florida Canals', *New York Times*, 17 May 1964.
8 'A Useful Anteater', *New York Tribune*, 11 July 1897.
9 'Monkeys as Helpers to Quadriplegics at Home', *New York Times*, 17 June 1987.
10 Richard Bulliet, *The Camel and the Wheel* (New York: Columbia University Press, 1990), p. 226.
11 'The Camel in Australia', *The Times*, 10 October 1894.
12 'Camels in Australia', *The Times*, 10 September 1884; 'The Camel in Australia', *Brisbane Courier*, 30 March 1907.
13 'Farming with Elephants: An Experiment in Africa', *The Times*, 8 March 1928.
14 D. Mota-Rojas et al., 'The Use of Draught Animals in Rural Labour', *Animals* 11:9 (2021), doi:10.3390/ani11092683.
15 Reinaldo Funes Monzote, 'Animal Labor and Protection in Cuba', in Martha Few and Zeb Tortorici (eds), *Centering Animals in Latin American History* (Durham, NC: Duke University Press, 2013), pp. 209–210.
16 Omar Inal, 'One-Humped History: The Camel as Historical Actor in the Late Ottoman Empire', *International Journal of Middle East Studies* 53 (2021), pp. 57–72.
17 Neil Pemberton, Julie-Marie Strange and Michael Worboys, *The Invention of the Modern Dog: Breed and Blood in Victorian Britain* (Baltimore: Johns Hopkins University Press, 2018), pp. 23–53; Chris Pearson, *Dogopolis: How Dogs and Humans Made Modern New York, London and Paris* (Chicago: University of Chicago Press, 2021), pp. 115–147.
18 Clay McShane and Joel A. Tarr, *Horse in the City: Living Machines in the Nineteenth Century* (Baltimore: Johns Hopkins University Press, 2007), p. 20.
19 McShane and Tarr, *Horse in the City*, p. 33.
20 McShane and Tarr, *Horse in the City*, pp. 2–7.
21 'Police Intelligence', *Morning Post*, 15 January 1841.
22 'Ilford Petty Sessions', *Morning Chronicle*, 8 July 1845.
23 'Training Dog Actors', *New York Times*, 5 October 1930.
24 On positive human–animal working relationships, see Jean Estebanez, Jocelyne Porcher and Juilie Douine, 'Are Screen Animals Actors?', and Nicolas Lainé, 'For a New Conservation Paradigm: Interspecies Labour: Examples from Human–Elephant Working Communities', both in Jocelyne Porcher and Jean Estabenez (eds), *Animal Labour: A New Perspective on Human–Animal Relations* (Bielefeld: Transcript Verlag, 2019), pp. 59–80 and 81–100, respectively.
25 'Training Dog Actors', *New York Times*, 5 October 1930.
26 Thomas Allsen, *The Royal Hunt in Eurasia* (Philadelphia: University of Pennsylvania Press, 2006), p. 79.

27 Neil Humphrey, 'Working Like a Dog: Canine Labour, Technological Unemployment, and Extinction in Industrialising England', *Environment and History* (2022), doi:10.3197/096734022X16384451127401.
28 'When the Elephant Goes on Strike', *Atlanta Constitution*, 29 November 1903.
29 Erica Fudge, *Quick Cattle and Dying Wishes: People and their Animals in Early Modern England* (Ithaca: Cornell University Press, 2018), pp. 116–119.
30 'Magawa the Mine-Sniffing Rat Ends Career in Cambodia on a High', *The Guardian*, 5 June 2021.
31 Tyler Parry and Charlton Yingling, 'Slave Hounds and Abolition in the Americas', *Past and Present* 246 (2020), pp. 69–108.
32 Bénédicte Boisseron, *Afro-Dog: Blackness and the Animal Question* (New York: Columbia University Press, 2018), pp. 37–80.
33 'Ilford Petty Sessions', *Morning Chronicle*, 8 July 1845; 'Police Intelligence', *Morning Post*, 15 January 1841.
34 Brian Fagan, *The Intimate Bond: How Animals Shaped Human History* (London: Bloomsbury 2015), pp. 103–130.
35 William Gervase Clarence Smith, 'The Donkey Trade in the Indian Ocean World in the Long Nineteenth Century', in Martha Chaiklin, Philip Gooding and Gwyn Campbell (eds), *Animal Trade Histories in the Indian Ocean World* (London: Palgrave Macmillan, 2020), pp. 147–179.
36 Samuel Graham Wilson, *Persian Life and Customs* (New York: Fleming, 1895), p. 54.
37 Peter Mitchell, *The Donkey in Human History: An Archaeological Perspective* (Oxford: Oxford University Press, 2018), p. 228.
38 Mitchell, *The Donkey in Human History*, pp. 205–217.
39 'Clever Donkeys', *Muswellbrook Chronicle*, 11 August 1900.
40 Mitchell, *The Donkey in Human History*, p. 44.
41 Songmeil Hu et al., 'From Pack Animals to Polo: Donkeys from the Ninth-Century Tang Tomb of an Elite Lady in Xi'an, China', *Antiquity* 94 (2020), pp. 455–472, https://doi.org/10.15184/aqy.2020.6.
42 Mitchell, *The Donkey in Human History*, pp. 83–84.
43 'On Behalf of the Egyptian Donkey', *The Times*, 16 January 1901.
44 'Donkeys Abandoned', *Queensland Times*, 7 June 1927.
45 Clarence Smith, 'The Donkey Trade in the Indian Ocean World', p. 149.
46 'Portsmouth Town Council', *Hampshire Telegraph*, 15 June 1840.
47 'Royal Society for the Prevention of Cruelty to Animals', *Morning Post*, 8 May 1850.
48 Hilda Kean, *Animal Rights: Political and Social Change in Britain since 1800* (London: Reaktion Books, 1998), p. 84.
49 'Bromyard', *Hereford Journal*, 20 May 1846.
50 'The Dog-Cart Nuisance', *Morning Post*, 4 July 1854.
51 'The Dog Cart Nuisance', *Morning Post*, 7 October 1846.
52 'Loughborough', *Leicestershire Mercury*, 2 June 1855.
53 '"Jobless" Elephants Demand Jobs', *The Times of India*, 20 August 1997.
54 'Dogs Parachuted 68 Miles', *The Times*, 10 December, 1956.
55 'Soviet Gains Data from Dog in Space', *New York Times*, 5 November 1957.
56 'Soviet Space Dogs on Show', *The Times*, 23 August 1960.
57 'Monkeys Back from Space', *The Times*, 29 May 1959.
58 'Space Chimpanzee is Safe after Soaring 420 Miles', *New York Times*, 1 February 1961.
59 'Space Chimp Home Safe!', *Chicago Daily Tribune*, 30 November 1961.

60 '55-Mile Ascent by Monkey', *The Times*, 5 November 1959; 'Space Chimp Home Safe!', *Chicago Daily Tribune*, 30 November 1961.
61 'Chimps Learn Space Roles', *Chicago Tribune*, 8 December 1963.
62 'Space Chimp Home Safe!', *Chicago Daily Tribune*, 30 November 1961.
63 'Chimps Trained for Space Flight', *New York Times*, 8 December 1963.
64 'Monkey Lost after Space Flight', *The Times*, 15 December 1959.
65 Amy Nelson, 'What the Dogs Did: Animal Agency in the Soviet Manned Space Flight Programme', *British Journal for the History of Science Themes* 2 (2017), p. 99.
66 'No Holiday for Ham', *Chicago Daily Tribune*, 2 February 1961; 'Space Chimpanzee Bars Encore', *New York Times*, 4 February 1961.
67 'Long Space Flight's Effect on Dogs', *The Times*, 17 May 1966; 'Bad Case of Nerves', *The Times*, 20 January 1962.
68 'Chimps Learn Space Roles', *Chicago Tribune*, 8 December 1963.
69 'US Air Force Plans to Retire Forgotten Space Race Veterans', *The Times*, 30 November 1995; Nelson, 'What the Dogs Did', p. 99.
70 Thomas R. Trautmann, *Elephants and Kings: An Environmental History* (Chicago: University of Chicago Press, 2015), pp. 259 and 244–246.
71 Terence D'Altroy, *The Incas, Second Edition* (London: Wiley Blackwell, 2015), p. 342.
72 'A Company of Elephants: Unique Army Unit in Burma', *The Times*, 28 November 1944.
73 Gervase Phillips, 'Animals in and at War', in Philip Howell and Hilda Kean (eds), *The Routledge Companion to Animal–Human History* (London: Routledge, 2018), p. 425.
74 John Lynch, *Simón Bolívar: A Life* (New Haven: Yale University Press, 2006), p. 73.
75 'Navy True to its Greatest Mascot', *New York Times*, 23 April 1910.
76 'Medal for Amethyst's Cat', *The Times*, 5 August 1949.
77 Phillips, 'Animals in and at War', p. 424.
78 Trautmann, *Elephants and Kings*, p. 19.
79 Fagan, *The Intimate Bond*, pp. 172–177.
80 Gervase Phillips, 'Writing Horses into American Civil War History', *War in History* 20:2 (2013), pp. 160–181.
81 John Miller, *Memoirs of General Miller* (London: Longman, 1829), Vol. I, p. 106.
82 James Hevia, *Animal Labour and Colonial Warfare* (Chicago: University of Chicago Press, 2018), pp. 27–49.
83 Sandra Swart, *Riding High: Horses, Humans and History in South Africa* (Johannesburg: Wits University Press, 2010), pp. 103–136.
84 'Navy True to its Greatest Mascot', *New York Times*, 23 April 1910.
85 'Soldiers' Dogs from Abroad', *The Animal World*, February 1919, pp. 18–19.
86 Hernán Cortés, *Hernan Cortés: Letters from Mexico*, trans. Anthony Pagden (New York: Grossman Publishers, 1971), p. 199.
87 Bernal Díaz del Castillo, *The True History of the Conquest of New Spain*, trans David Carrasco (Albuquerque, University of New Mexico Press, 2008), p. 50.
88 Díaz del Castillo, *The True History of the Conquest of New Spain*, pp. 51–52.
89 Díaz del Castillo, *The True History of the Conquest of New Spain*, pp. 58–59.
90 Díaz del Castillo, *The True History of the Conquest of New Spain*, pp. 40–41.
91 Cortés, *Letters from Mexico*, p. 252.
92 Díaz del Castillo, *The True History of the Conquest of New Spain*, p. 50.
93 Cortés, *Letters from Mexico*, p. 141.

94 Mackenzie Cooley, *The Perfection of Nature: Animals, Breeding and Race in the Renaissance* (Chicago: University of Chicago Press, 2022), p. 115.
95 Díaz del Castillo, *The True History of the Conquest of New Spain*, p. 58.
96 Alex Hidalgo, *Trail of Footprints: A History of Indigenous Maps from Viceregal Mexico* (Austin: University of Texas Press, 2019), p. 11.
97 Cortés, *Letters from Mexico*, p. 249.
98 Díaz del Castillo, *The True History of the Conquest of New Spain*, p. 108.
99 Bernardino de Sahagún, *General History of the Things of New Spain by Fray Bernardino de Sahagún: The Florentine Codex* (Place of publication unknown, 1577), Book 12, p. 961.
100 Isabelle Schürch, 'Liminal Lives in the New World', in Clemmens Wischermann, Aline Steinbrecher and Philip Howell (eds), *Animal History in the Modern City: Exploring Liminality* (London: Bloomsbury, 2019), pp. 32–33.
101 Rachael Pasierowska, 'Atlantic History from the Saddle', in Chiara Mengozzi (ed.), *Outside the Anthropological Machine: Crossing the Human–Animal Divide and Other Exit Strategies* (New York: Routledge, 2020), p. 62.
102 Annie Martin, *Home Life on an Ostrich Farm* (London: George Philip and Son, 1890), pp. 242–243.

Further Reading

Boisseron, Bénédicte, *Afro-Dog: Blackness and the Animal Question* (New York: Columbia University Press, 2018)

Bulliet, Richard, *The Camel and the Wheel* (New York: Columbia University Press, 1990)

Clarence Smith, William Gervase, 'The Donkey Trade in the Indian Ocean World in the Long Nineteenth Century', in Martha Chaiklin, Philip Gooding and Gwyn Campbell (eds), *Animal Trade Histories in the Indian Ocean World* (London: Palgrave Macmillan, 2020), pp. 147–180

Fagan, Brian, *The Intimate Bond: How Animals Shaped Human History* (London: Bloomsbury 2015)

Fudge, Erica, *Quick Cattle and Dying Wishes: People and their Animals in Early Modern England* (Ithaca: Cornell University Press, 2018)

Funes Monzote, Reinaldo, 'Animal Labor and Protection in Cuba', in Martha Few and Zeb Tortorici (eds), *Centering Animals in Latin American History* (Durham, NC: Duke University Press, 2013), pp. 209–210

Hevia, James, *Animal Labour and Colonial Warfare* (Chicago: University of Chicago Press, 2018)

Humphrey, Neil, 'Working Like a Dog: Canine Labour, Technological Unemployment, and Extinction in Industrialising England', *Environment and History* (2022), DOI: 10.3197/096734022X16384451127401.

Inal, Omar, 'One-Humped History: The Camel as Historical Actor in the Late Ottoman Empire', *International Journal of Middle East Studies* 53 (2021), pp. 57–72

McShane, Clay, and Joel Tarr, *Horse in the City: Living Machines in the Nineteenth Century* (Baltimore: Johns Hopkins University Press, 2007)

Mikhail, Alan, *The Animal in Ottoman Egypt* (Oxford: Oxford University Press, 2014)

Mitchell, Peter, *The Donkey in Human History: An Archaeological Perspective* (Oxford: Oxford University Press, 2018)

Nelson, Amy, 'What the Dogs Did: Animal Agency in the Soviet Manned Space Flight Programme', *British Journal for the History of Science Themes* 2 (2017), pp. 79–99

Phillips, Gervase, 'Writing Horses into American Civil War History', *War in History* 20:2 (2013), pp. 160–181

Phillips, Gervase, 'Animals in and at War', in Philip Howell and Hilda Kean (eds), *The Routledge Companion to Animal–Human History*, (London: Routledge, 2018), pp. 422–445

Parry, Tyler, and Charlton Yingling, 'Slave Hounds and Abolition in the Americas', *Past and Present* 246 (2020), pp. 69–108

Porcher, Jocelyne, and Jean Estabenez (eds), *Animal Labour: A New Perspective on Human–animal Relations* (Bielefeld: Transcript Verlag, 2019)

Saha, Jonathan, *Colonizing Animals: Interspecies Empire in Myanmar* (Cambridge: Cambridge University Press, 2022)

Shepherd, Verene, 'Livestock and Sugar: Aspects of Jamaica's Agricultural Development from the Late Seventeenth to the Early Nineteenth Century', *The Historical Journal* 34:3 (1991), 627–643

Swart, Sandra, *Riding High: Horses, Humans and History in South Africa* (Johannesburg: Wits University Press, 2010)

Trautmann, Thomas, *Elephants and Kings: An Environmental History* (Chicago: University of Chicago Press, 2015)

3   Companionship

As well as keeping animals for food and labour, humans have also kept them for companionship. These animals form part of a special, privileged category, which we know today as pets. Historians have provided various definitions of pets. Keith Thomas claims that what differentiates pets from other animals is that they are named, allowed inside the house and never eaten.[1] Katherine Grier expands on this definition, noting that pets were singled out for special attention designed to promote their well-being; they were often referred to as 'favourites' in Western cultures and are still treated that way. The actual word 'pet' most likely emanates from the French 'petit' ('small'), and was applied first to spoiled children, later to animal companions.[2] Above all, what distinguishes the human–pet relationship from other human–animal relationships is that it is based on an emotional bond, rather than any utilitarian need – though, as we shall see, the lines between pet and labouring animal could be blurred.

In recent years, pet keeping has attracted a growing number of studies. Historians have analysed the power relations between owners and their pets. They have explored the role of pets as cultural signifiers and they have examined prevailing social critiques of pet keeping. They have also asked how pet keeping has evolved over time, and when it first attained its modern form. What constituted a pet? Who owned pets and why? Did animals benefit from their status as companion species? While most scholarship on pet-keeping has focused on Europe and North America, recent studies of pets in such diverse settings as eighteenth-century Mexico and twentieth-century Japan are gradually expanding our knowledge of pets in non-Western contexts, allowing us to gain a clearer picture of the role animal companions have played in world history.

One reason why pets have attracted so much attention from historians is that they have left more extensive traces than other types of animals. Because of their close physical and emotional proximity to their human owners, pets have frequently been written about and pictured, appearing in diaries, letters, newspaper articles, paintings and photographs. In a few cases, animal

DOI: 10.4324/9781003181996-4

companions have left more idiosyncratic traces, from cages and collars to taxidermy mounts; a pet cat at a monastery at Deventer in the Netherlands urinated on a medieval manuscript, prompting the exasperated monk to draw a picture of it.[3] By drawing on this eclectic range of sources, historians have been able to probe the human–animal relationship more deeply and uncover complex cross-species interactions that were not simply about exploiting animals for their meat or muscle. Though a minority group within the animal kingdom, pets therefore generate a disproportionate amount of source material and provide a valuable window onto human–animal relationships in the past.

Pets, Ancient and Modern

When did humans first start to keep pets? Is pet keeping a modern phenomenon, or a practice with much older roots? Who has owned pets, and for what purpose? To what extent has the role of animal companions changed over time?

Pet keeping was common in the ancient world. The Egyptians venerated cats, adorning them with jewellery and shaving their eyebrows when a beloved feline died. The Greeks and Romans also kept a variety of pets, ranging from dogs, hares and goats to nightingales, ravens and parrots. Pliny offers a description of the latter in his *Natural History* (written 77AD) noting that they were imported from India, became 'especially frolicsome under the influence of wine' and could be taught to talk by beating them over the head with an iron bar.[4] While it is difficult to know the exact relationship that existed between humans and pets in the pre-modern period, there is strong evidence that many were valued highly. Some animal companions appeared on mosaics or in hieroglyphs. Others were the subjects of poignant epitaphs when they died. Yet others were accorded formal burials. In 2011, for example, archaeologists at Berenike on the Red Sea coast of Egypt excavated the skeletons of eighty-six cats, nine dogs and several monkeys dating back to the second century AD, some of which wore ostrich shell beads and iron collars around their necks. Since the animals were not mummified, and had, in some cases, reached old age, researchers believe that they were probably pets.[5]

In the medieval and early modern periods, pet keeping was largely the preserve of the elite, who could afford to devote time and money to animals with no useful function. Clerics and nuns kept pets in monasteries and abbeys. Aristocrats and, later, members of the urban middle classes, accumulated pets as a form of conspicuous consumption, while kings, queens and princesses kept animal companions in their palaces. Isabella d'Este Marchioness of Mantua (1474–1539) owned a small cat, or 'animalino', which she carried about in her sleeve.[6] Catherine the Great of Russia

(1729–1796) recounted 'giving hazelnuts to a white squirrel whom I tamed myself' and visiting 'a charming monkey of mine who is so mad I can never see him without laughing'.[7] Valued for their rarity, beauty and ability to entertain, pets provided company for lonely princesses and served as welcome distractions from the stultifying etiquette of court life. Among the upwardly mobile bourgeoisie, meanwhile, domestic animals functioned as symbols of prosperity, appearing in family portraits alongside other luxury items. A portrait of Jan Jacobszoon Hinlopen from 1662, for example, shows the wealthy Dutch clothier flanked by his wife and children, all attired in fashionable silks and surrounded by several domestic animals, including a squabbling dog and tabby cat, a spaniel begging for a biscuit and a colourful species of New World bird, probably a parrot (Figure 3.1). The latter was a product of Europe's burgeoning trade with Africa, Asia and the Americas, and its presence in Hinlopen's home a testament to the mercantile strength of the Dutch Republic (1588–1795).

The nineteenth century witnessed the extension of pet keeping to the industrial bourgeoisie, and, increasingly, to many working-class people. Faster shipping and railroad expansion facilitated the importation of more exotic species at lower prices, making even non-native animals such as monkeys and parrots accessible to the middle and lower classes. Some species – notably canaries and goldfish – started to be bred explicitly for the pet trade in specialised facilities from the 1870s, further lowering their respective prices, while exotic breeds of dogs and cats such as Pekingese dogs and Siamese and Persian cats gained increasing popularity in the West, where

Figure 3.1 Gabriel Metsu, *Portrait of Jan Jacobsz Hinlopen and his Family*, 1662, bpk / Gemäldegalerie, SMB / Jörg P. Anders.

Source: bpk / Gemäldegalerie, SMB / Jörg P. Anders.

they were imported in large numbers; in 1884, for instance a visitor to Charles Jamrach's wild animal emporium in London encountered an Afghan hound with 'the head and general build of a greyhound' ears 'long' and 'silky' like those of a setter and similar 'tufts of hair' on the 'surface of his paws'.[8] The nineteenth century also saw an increasing commercialisation of pet-keeping, with pet-related accessories advertised and retailed in ever growing quantities. Spratt's patented the first dog biscuits in 1860, giving rise to a market in specialised pet foods. Pet manuals were published on the care of a wide variety of species, while ornamental bird cages, collars, jackets and even shoes for dogs were sold in boutiques and, later, department stores, allowing owners to dress their pets in the latest fashions.[9] Writing in 1896, journalist Guy Tomel estimated that there were 'not less than 5,000 dogs in Paris who have their clothes made to order', with the typical canine wardrobe including 'a set of night gowns in batiste or in silk', 'half a dozen embroidered handkerchiefs', shoes made of 'anhydrous leather or India rubber', a 'travelling coat' made of ermine or sealskin and a white coat and 'little sailor hat' for walks on the beach.[10] The breeding of dogs (and, to a lesser extent, cats) for form over function also began in earnest in the mid-nineteenth century, giving rise to official breed standards and, from the 1860s and 70s, to the first dog and cat shows.[11]

Many of these trends have continued into the twentieth and twenty-first centuries, as pet keeping has become increasingly widespread and commercialised. Pets remain extremely popular in the West, where cheaper and more nutritious pet food and the expansion of small animal veterinary practices have improved pet well-being and survival rates. The emergence of an urbanised and affluent middle class in Asia has led simultaneously to significant growth in the number of household pets – and the accessories that go with them. In post-war Japan, for example, dog-keeping became increasingly popular, giving rise, by the early twenty-first century, to 'a flourishing industry of pet-sitting services, fee-charging dog parks, pet-training schools, pet hotels, pet cafes, beauty salons for pets and pet cemeteries'.[12] In China, pet keeping – condemned as a bourgeois luxury under Mao – has expanded dramatically since 1978, with an estimated 168 million dogs in urban and rural households by 2017.[13] While some recent developments have been positive for pets, improving their health and quality of life, others have been detrimental. Continued inbreeding to meet artificial breed standards has impacted negatively on the well-being of many dogs, while neglect and abandonment remain rife. The trade in exotic pets, moreover, has reached record levels, decimating wild animal populations and damaging local ecosystems.[14] In Florida, for example, Burmese pythons, originally imported as pets, have colonised the Everglades, breeding at a rapid rate and steadily devouring the local fauna – including alligators![15] The twentieth century has thus seen the increasing globalisation of pet-keeping, as exotic species travel between continents and Western forms of animal companionship are exported to modernising societies in Asia, Africa and Latin America.

Dogs, Cats … or Hedgehogs?

While pet keeping has a long pedigree, the types of animals chosen as companions have varied according to time and place, reflecting shifting cultural preferences. Dogs have been popular across a broad range of cultures, though they have often occupied a liminal position between labouring animal and pet. Cats have fluctuated in popularity over time, venerated by the ancient Egyptians, but associated with witchcraft in early modern Europe. Canaries (from the Canary Islands) were domesticated in Europe in the sixteenth century and bred in large numbers in the nineteenth century, primarily in Italy and Germany. Guinea pigs were first introduced to Europe as pets in the sixteenth century, following the Spanish conquest of Peru (where they were reared for food and ritual sacrifice), while hamsters only appeared in pet stores in the 1940s, having originally been imported and bred as experimental animals for scientific research.[16] Other species that were once popular pets have since fallen out of favour – at least in the West. Squirrels and hedgehogs, for example, were commonly kept as pets in early modern and nineteenth-century Europe and North America, but are now rarely kept as companions.[17] John Singleton Copley's portrait of Daniel Crommelin Verplanck (1771) depicts the nine-year-old Daniel holding a grey squirrel with a gold collar and chain (Figure 3.2).

Exotic animals such as parrots, monkeys and tortoises have also been kept as pets, especially among the privileged elite. Parrots were prized for their ability to speak. Tortoises were believed (wrongly) to eat beetles, so were kept by the Victorians partly as a form of pest control. Monkeys were valued for their capacity to entertain, and often sold on this basis.[18] In 1900 one vendor in *The Bazaar* advertised a 'Lovely monkey, 2 years old, very tame, household pet, boxes with little boy, ducks and feints like a professional, pure white cat sleeps in its arms'.[19] Other even more unusual species have also served as domestic companions from time to time, in most cases brought back by travellers or sourced through large-scale animal dealers such as Charles Jamrach in London or Carl Hagenbeck in Hamburg. In 1814 British museum proprietor William Bullock kept a vampire bat, who 'was fond of white wine, of which it would take nearly half a glass full at a time' and 'slept suspended, with its head downwards, wrapping its soft wings round its body in the form of a mantle'.[20] In 1935 thirteen-year-old Elma Barnes from Inverloch, Victoria, owned a pet wallaby called Betty who ate 'anything it sees – pine needles, sticks, butter, grass, meat, cakes and paper', and cuddled up in a box between her two cats.[21] In 1954 Lilo Hess from Pennsylvania had a pet anteater named Teddy, who 'delighted in licking things' (including the

Figure 3.2 John Singleton Copley, *Daniel Crommelin Verplanck* (1771).
Source: Heritage Image Partnership Ltd / Alamy Stock Photo.

cat), 'loved to be petted and have his head scratched' and 'responded to his name' when called (Figure 3.3).[22] In 1963 zoologist Captain J.E. Edwards owned a 'four-foot alligator named Trudy' who travelled with him 'on the back seat of his car'.[23] Pet status was thus conferred upon a wide range of species – though dogs and cats were (and are) more amenable to the demands of domesticity than alligators and anteaters.

While species has been a major determinant of pet status, it should be noted that this status is mutable, and that the line between cherished companion and working animal/food is often blurred. This complicates the definition of 'pet' and highlights the degree to which it is culturally constructed. First, it is important to recognise that while it is common to classify some species as domestic pets and others as farm animals or wild animals, these boundaries can be quite fluid. Stray dogs, for instance, unsettle the boundary between wild and tame, and have often been culled as possible vectors of rabies – initially in Europe and North America, later in colonial settings such as Singapore and Nigeria.[24] White mice and rats, specially bred to bring out their albinism, are seen as pets – especially for children, but their wild

ON the preceding page we describe how Miss Lilo Hess bought *Teddy*, the anteater, and took him to her home in Pennsylvania to be a playmate for her baby chimpanzee, *Christine*. Miss Hess recalls the first meeting between the two animals when the anteater walked slowly out of his crate and the chimp, fascinated, tried to pet the new toy, which hit back, and the chimp only just ducked in time. When Miss Hess tried to touch *Teddy*, he lifted his fore-leg to strike at her too, so, after giving him some food, she retreated, to let him settle down in his new surroundings. A cage was arranged for him in the porch, and a large pen out of doors. After sleeping for the rest of the day and all night, *Teddy* was more at ease and came out to eat his food and let Miss Hess pet him but. Miss Hess says : " His front leg would rise ready to strike, but it remained only a gesture. He never quite lost this habit, even though he never really struck. It was like a reflex action." A collar was put round *Teddy*'s middle, which he always kept on and did not seem to mind at all. He liked best to be tied to a tree in the grass so that he could indulge in his overriding passion, which was, naturally, ant-eating. He preferred the tiny black ants to the red ones, and picked out the larvæ and eggs first before concentrating on the ants. The chimpanzee and the anteater never became close friends, as *Teddy* never trusted the chimp, whose quick movements seemed to make him nervous and upset.

(*Continued below, left*.)

Continued.]
be petted and have his head scratched by Miss Hess and would stand very still and close his eyes. The only thing *Christine* consented to was letting *Teddy* pull her along in a little waggon. But she soon got impatient, since he stopped all the time to look for ants. When he sometimes turned to sniff her, she would jump away as fast as she could. *Teddy* delighted in licking things, and Miss Hess comments that " after touching the anteater's tongue many times, it seems to me that the tongue itself is not sticky but the saliva is. He would run his tongue with great speed back and forth over my hand, and there was an adhesive coating on it afterwards." *Teddy* licked the broom and the chairs and even the cat, and was very much interested in the kitchen. Miss Hess thinks that he might have smelt ants in it, as she had had an invasion of them in the spring. However, they had since departed and *Teddy* hardly ever found any, but he just kept on looking. Miss Hess says : " There were no signs of great intelligence, but he was not stupid . . . he learned to fit himself into our household in a very short time. He knew my voice and responded to his

name . . . When I came in the morning to put his leash on, he would be as docile as a lamb. He would come towards me and stand very still so I could fasten it. But in the evening it was a different story. He knew I had come to take him away from his ants, and he would rebel as best he could. . . . He never really struck out at me though, and I would tuck him under my arm and carry him in. Eventually)

(*Continued above, right*)

Figure 3.3 'An Unusual Family Pet in a Pennsylvanian Home: Teddy the Giant Anteater', *The Illustrated London News*, 13 March 1954, p. 406.

Source: © Illustrated London News Ltd/Mary Evans.

counterparts have often been viewed as pests to be exterminated. Most cows, pigs and chickens are reared for food, but individuals have often been selected as pets and accorded similar treatment to dogs and cats – though they frequently end up being slaughtered when they reach maturity. A manual entitled *The Guinea Pig, or Domestic Cavy, for Food, Fur and Fancy* (1886) advocated breeding guinea pigs for show and as pets, but suggested that lower-quality offspring could be eaten in cavy ragouts, cavy curries and cavy pies (for which it provided recipes).[25] All of these cases trouble the parameters of pet status and expose some of the inconsistencies in humans' treatment of other animals.

Second, it should be noted that an individual animal can experience different treatment at different periods of its life, moving between the categories cited above. Dogs and cats that were once pets can be abandoned by their owners and become strays. Pet horses in the nineteenth and early twentieth centuries could be named and petted while they remained useful, but sold to the knacker's yard when they grew old – a fate they often shared with cossetted pigs, chickens and the occasional pet rabbit. Cats and dogs kept for hunting, conversely, might nonetheless be named, allowed in the house and, if they were lucky, kept on in the household after they ceased to work. Pet kangaroos (popular in twentieth-century rural Australia), had a habit of escaping and returning to the wild, and were sometimes confused with their wild cousins; Pat Wells's pet kangaroo, Joe, was shot and blinded in one eye by a passing sportsman in the 1930s, despite wearing a 'red collar' to identify him as a companion animal.[26] Once again, this poses significant ethical questions about what is, and is not, a pet, and what rights should be accorded to animals classified as companions. 'Pet' is not therefore a stable category; animals originally designated as pets could be made to work, killed and even eaten, while wild or feral animals could occasionally achieve pet status.

Lastly, it is important to recognise that some non-Western cultures have taken a different approach to pet keeping, producing more complex human–animal relationships. While Western pet keeping has generally focused on mammals and birds, for example, in Japan children have long kept insects (*mushi*) for education and amusement. According to Erick Laurent, who has studied the keeping of *mushi* in past and present Japan, 'The first reference … to the selling of "autumn singing insects" (crickets) seems to date back to 1685 in Kyoto', when the insects were 'sold in big square baskets suspended by a pole put on the shoulders … on top of which were placed smaller cages sold together with the *mushi*'. These creatures (most commonly crickets, cicadas, rhinoceros beetles, dragonflies and spiders) lived in the home and were often bred, though not usually for food. While there is some debate as to whether *mushi* qualify as pets or not in the Western sense – they only live a short time and are not usually mourned when they die – Laurent argues that, on balance, *mushi* should be classed as pets, since they are treated more like playmates than toys and participate in 'a special relationship ... involving pleasure and play'.[27] In Japan, therefore, the definition of pet extends beyond mammals, birds and fish to encompass insects – a class of animals traditionally excluded as companions in the West (although stick insects and tarantulas (technically arachnids) have become popular in recent years).

Pet-keeping practices in Amazonia likewise challenge Western assumptions about what constitutes a pet. We tend to assume that pet status is determined by species (dogs, for example, are pets, pigs and cows are not), but this is not true everywhere. In her study of human-animal relations in Pre-Columbian and colonial Spanish America, Marcy Norton describes how lowland South American Indians kept some members of a particular species as pets (for instance parrots, monkeys, tapirs), but readily ate their wild counterparts. Indeed, pets were usually brought back by hunters and reared by their wives, who pre-masticated food for fledgelings or breast-fed mammals. A cacique (chief) in early-sixteenth-century Hispaniola, for example, tamed a young manatee that became trapped in his fishing nets, keeping the animal in a large lake, naming him Matu (which means 'generous' or 'noble') and feeding him on yucca, cassava and millet.[28] An indigenous family in nineteenth-century Brazil owned a young uakari (scarlet-faced) monkey, rearing the 'frisky little fellow...in the home amongst the children' and allowing him 'to run freely and take [his] meals with the rest of the household'.[29] Norton concludes that we should not apply European conceptions of pet keeping and farming to non-European societies and emphasises that other paradigms for relating to animals may exist. As she remarks:

> The inhabitants of Europe and lowland South America were accustomed to and comfortable with the idea of eating animals. However, they developed different solutions to reconcile this fact with their awareness that they could and did form affective relationships with some of these beings. The European solution was to prohibit eating certain beings and objectify most of these beings whose consumption was licit. The South American solution was to classify beings according to the contingent condition of their tameness. While livestock husbandry was most fundamentally about killing and eating animals who were fed, familiarization was predicated on the belief and practice that those who were fed were kin, and not, therefore, to be killed and eaten.[30]

So while Europeans have tended to class certain species as pets, some Amazonian cultures have designated individual animals as pets, regardless of species, due to the specific relationship they have forged with them.

Pets and Society

Pet-keeping has taken different forms in different societies. It has also, in the process, posed important social questions, acting as a barometer for shifting attitudes towards other species. Pets have been paraded as status symbols, cherished as surrogate children and (occasionally) subjected to ritualised violence. Owing to their proximity to their human owners, they have also been summoned as metaphors for broader social problems, featuring regularly in debates about religious orthodoxy, good governance and appropriate female conduct. What, therefore, can pet keeping tell us about the priorities and preoccupations of different human societies? What criticisms have been directed at pets and the people who own them? In what ways have pets become metaphors for wider social ills?

Much critical commentary on pet-keeping has centred on its perceived links to luxury and dissipation. By definition animals with no practical purpose, pets have often been seen as a distraction from more worthy purchases or pursuits, and, when their owners were female, as a frivolous diversion from marriage and child-bearing. Writing in 1570, John Caius denounced lapdogs as 'instruments of folly for [wanton women] to play and dally withal, to tryfle away the treasure of time'.[31] Two and a half centuries later, in 1817, another British author, John Rippingham, delivered a similarly stern rebuke to pug-loving 'English females', remarking that it was 'difficult to believe that any lady should afford such proof against herself that she could find no amusement more pleasing or respectable than that of nursing a dog – an employment from which she could derive no information, nor obtain the smallest share of praise or credit'.[32] The reasons why such behaviour was condemned varied over time – at least superficially. In early modern England elderly unmarried women who kept dogs, cats and other animals were often accused of engaging in witchcraft. In eighteenth-century Europe women who kept pets rather than marrying and having children were seen as reneging on their patriotic duty to procreate, in an era when many rulers viewed population growth as essential to the creation of strong states with large armies.[33] In nineteenth-century Britain, meanwhile, owning multiple pets (what we would now call hoarding) was viewed as a sign of eccentricity, and, possibly, a danger to public health; in 1867 Mr J. Noakes, 'an inspector of nuisances', filed a complaint against a young female artist named Miss Deen, whose residence housed from '100 to 200' cats, 'all in an unhealthy state and exceedingly dirty'.[34] Underlying all of these concerns, of course, was a misogynistic disdain for childless women and a belief that pet ownership – especially of cats and lapdogs – denoted selfishness and ignorance. Though less overtly expressed, such anxieties persist into the twenty-first century, whether in the Western stereotype of the 'mad cat lady' or the growing concern in some Far Eastern countries about the declining birth-rate.

In twenty-first-century Taiwan, for instance, many young people are choosing to own dogs and cats rather than have children, with pets now outnumbering children under the age of fifteen.[35]

While it was primarily women who stood accused of lavishing excessive attention on pets, men have also come in for criticism on this count, especially when their devotion to animals impaired their ability to govern. The French King Henry III (1574–1589), for example, was mocked by his contemporaries for surrounding himself with a breed of small dogs known as lion dogs (they were bred in Lyon) and rejecting more stereotypically 'noble' animals such as lions. Contemporaries perceived this as synonymous with Henry's fraternisation with effeminate male favourites, known as *Mignons*, and alleged that the monarch's passion for cossetted canines was eating into the royal finances; every eight dogs purportedly had their own governess, servant and packhorse to carry them.[36] The case of King George IV's giraffe offers another revealing example of how culturally inappropriate relationships with animals could provide fodder for political opponents. A gift to George from the Pasha of Egypt, the giraffe arrived in England in 1827 and was visited regularly by the king, who clothed the animal in a blanket to protect it from the cold and summoned surgeons to treat its swollen knee.[37] While such attentions might seem touching to the modern reader, contemporaries were less impressed, subjecting George to a barrage of ridicule. One critic, Sir Henry Halford, made a connection between the giraffe's ailments and George's own famed obesity, informing the Privy Council that 'the indisposition of the Giraffe at Windsor has arisen from the animal's loyal sympathy in his Majesty's twinges in his toe, in his late fit of gout'.[38] Another, a destitute merchant, wrote an acerbic letter to the *Liverpool Mercury* in which he accused George of putting the care of his exotic pet above the suffering of his subjects:

> I have perpetually before me the afflicting sight of a wife and six daughters almost heart-broken, not only deprived of the comforts of life ... but almost destitute of its necessaries, and that, too, without any prospect of amendment. But this is nothing compared with the misery I have undergone from solicitude for 'that rare animal, the giraffe', which at present appears very properly to occupy most of his Majesty's attention.[39]

A series of satirical cartoons also circulated widely, showing the king riding on the giraffe's back with his mistress, coddling it in his boudoir and raising up the ailing animal with a specially designed pulley. Like Henry III's lapdogs, George IV's giraffe thus became the focus of opposition for the king's many detractors, and a convenient vehicle through which to attack an already unpopular monarch.

Pets could also cause controversy in the spiritual realm when they transgressed established conventions or breached religious taboos. Though often treated as honorary humans, pets were not accorded human status in scripture and their participation in certain religious rights was condemned as sacrilegious. In Victorian Britain, lengthy debates took place over whether pet dogs had souls and might join their owners in the afterlife.[40] In eighteenth-century Mexico, meanwhile, several elite dog-owners violated Catholic dogma by subjecting their canine companions to imitation baptisms and marriages, inciting the wrath of the Inquisition (the reason we know about these ceremonies is because they were recorded in inquisitorial archives). On one occasion, several individuals from Mexico City were hauled before the Inquisition for taking part in a marriage ceremony for two dogs. On another, a tailor named Joseph Armas was arrested after he conducted a mock baptism of two dogs, putting salt on the animals' paws and immersing their bodies in water. Canine marriages and baptisms were deemed immoral because they subverted both the sanctity of the holy sacraments and the traditional relationship between humans and animals. As Zeb Tortorici explains: 'In canine weddings and baptisms, carnivalesque religious rituals veered dangerously close to heresy, but the larger issue at stake was that, at least in the eyes of ecclesiastical authorities, these acts challenged the divinely ordained natural and social orders'.[41] Like Henri III's excessive affection for his lapdogs, the marriage of two dogs undermined and mocked traditional social hierarchies, blurring the boundaries between human and animal.

Finally, the fact that pets suffered a loss of liberty invited criticism from humanitarians and social reformers, who often drew parallels with prisoners, servants and slaves. During the French Revolution, for instance, the image of the caged bird was deployed by opponents of monarchical despotism to illustrate the fate of prisoners in the Bastille.[42] The act of domestication, meanwhile, was often equated with enslavement, especially when this was done purely for amusement; one early twentieth-century writer compared the shipment of grey parrots to the 'horrors of the "middle passage" in the old days of the slave trade to America and the West Indies', so extreme was the suffering and so high the level of attrition.[43] As Yi-Fu Tuan has shown, moreover, there has often been a deeply uncomfortable parallel between the treatment of animals and some subaltern groups of human beings, who have been viewed in like manner as exotic extravagances and objects of conspicuous consumption. In Britain between the sixteenth and eighteenth centuries, it was common for rich men and women to keep African page boys, who attended their masters in colourful liveries. In many medieval and early modern courts, fools and dwarves were conscripted as a source of amusement, sometimes treated with indulgence, but simultaneously consigned to a state of perpetual childhood. These individuals were regarded, in effect, as human pets, and often pictured with captive animals, reinforcing this disturbing parallel.[44] Pet keeping thus became enmeshed in wider social debates about freedom, exploitation and oppression and served as a metaphor for imprisonment, enslavement and other forms of human servitude.

Dominance and Affection

Dominance

While some studies of pet keeping have focused primarily on what pets reveal about the people that own them, a second strand of research concerns the power dynamics of the human–pet relationship, and how this has been experienced by the animals themselves. What kind of a relationship have humans had with their pets? Is it one based on dominance and exploitation, or one founded in love and affection? Have pets been viewed primarily as cherished companions, or as luxury commodities? Has the balance between dominance and affection shifted over time, or remained fairly constant? How far, if at all, has the keeping of pets influenced humans' treatment of other animals, either wild or domestic? An analysis of the experiences of pets at different points in their lifecycle presents a complex picture, with instances of both callous neglect and genuine intimacy.

If we begin by considering the sourcing, breeding and retailing of pets, the 'dominance' side of the equation certainly seems to predominate, particularly in the case of exotic animals. Taken from the wild by hunters and trappers, species such as parrots, monkeys and tortoises were forced to endure long, harrowing transcontinental and transoceanic voyages to dealers in distant countries, succumbing in large numbers to trauma, inappropriate diets and disease. Writing in 1867, when the exotic animal trade was reaching its peak, naturalist Frank Buckland, recounted how his pet monkey, Susey, arrived at Charles Jamrach's exotic animal emporium 'lying on her side breathing very hard, and very, very ill', only surviving through the prompt administration of 'port wine, beef-tea and hot flannels'.[45] Ten years later, *The Bazaar*'s bird expert, C.W. Gedney, claimed that 90% of African grey parrots taken from the African interior died from consumption (lung disease) soon after they reached Britain.[46] His successor, Dr Greene, who performed post-mortems on readers' pets, informed one grieving owner that 'Not one [grey parrot] survives out of a thousand, and yet people will keep on wasting their money in buying them. It is not a bit of use your buying another, as it is bound to go the same way'.[47] The exotic pet trade thus exacted a terrible toll on wild animals, decimating populations and inflicting great cruelty on trafficked individuals. Though more strictly regulated by import and export bans since the late twentieth century, commerce in exotic pets continues to pose a threat to many endangered species, pushing some to the brink of extinction.[48] A 2015 study reported that hundreds of slow lorises were being smuggled illegally out of India, southern China and south-east Asia to meet a growing demand for pet lorises in Japan, where the primates sell for up to £5,800 each.[49]

More common animal companions such as dogs, cats and guinea pigs are not generally taken from the wild but bred specifically for the pet trade (though in eighteenth- and nineteenth-century Europe there was a strong demand for wild-caught native species such as songbirds, squirrels and hedgehogs).[50] Before 1900, this was often an ad hoc process, performed on a local level; most people sourced their cats and dogs from friends and relatives whose animals had given birth to unwanted offspring. Since the early twentieth century, however, largescale breeding of pets has become the norm, with significant welfare implications. Small animals such as hamsters, budgerigars and tropical fish have been bred in captivity since the 1920s and sold in pet stores across the globe, while large number of dogs have been reared on puppy farms since the 1940s, often in appalling conditions. The advent of breed in the mid-nineteenth century, moreover, has spawned a desire for designer dogs, selected for specific physical features rather than to perform a specific task (e.g. game retrieval, sheep herding), giving rise to deformities related to inbreeding. Bulldogs, for example, have metamorphosed from lithe, muscular animals with strong jaws into the short, squat dogs we know today, with drooping jowls, shortened legs and flattened noses, through which they often struggle to breathe. The supply of pets therefore involves a good deal of animal suffering – much of it unseen by the consumer – and entails the reshaping of animal bodies to satisfy human caprices. In twenty-first-century China it is even possible (and legal) to clone your pet (for an average cost of £28,000), a practice that raises important questions about the value of companion animals as individuals.[51]

Once acquired or purchased by owners, the lives of many pets may have improved, but others have continued to suffer from either chronic mistreatment or active abuse. At the more severe end of the scale, pets have been kept in unsuitable cages, fed inappropriate food or deprived of exercise and stimulation, impacting their well-being and shortening their lives. Writing in 1893, for instance, one reader of *The Animal World* alleged that parrots were 'often placed in bell-shaped cages, so narrow that I can scarcely think that they are even able to expand their wings to their fullest extent'.[52] At the more benign end of the spectrum, pets have been (and are) flaunted as status symbols, forced to moderate their natural behaviours and trained to perform tricks for their owners' entertainment, suggesting again that dominance trumps affection. In eighteenth-century Spain, for example, parrots at the royal court were fed 'wine soup' to make them talk.[53] In nineteenth-century Burma, pet owner H.C. insisted on dressing her pet gibbon, Dinah, in a 'pale pink silk dress trimmed with lace, and a hat', even though the ape clearly resented the treatment, and 'will allow no one to perform the operation but me, savagely biting if they attempt it'.[54] Even loved pets could thus experience pressure to conform to human priorities, and many have had to endure training, petting or (in extreme cases) surgical intervention in order to do so.

Lastly, of course, pets, like all animals, could be the victims of deliberate abuse, suffering neglect, starvation or even physical violence. Old, sick or unwanted animals were often abandoned by their owners and left to fend for themselves. Cats were left unfed when their owners went on holiday (prompting an RSPCA campaign against the practice in Britain in the 1890s), while, before the advent of spaying in the twentieth century, most kittens were drowned at birth to keep domestic feline populations under control.[55] Some pets also experienced more extreme forms of abuse at the hands of sadistic or quick-tempered humans, from teasing to brutal beatings. In 1880s South Africa the owners of a tame baboon called Sarah subjected her to a series of 'unkind practical joke[s]', tricking her into eating 'two ... succulent slices of pumpkins [on which] cruel hands had spread a thick layer of mustard' and hiding a dead snake between paper for her to unwrap – causing her to 'swoon'.[56] In 1936 the American Society for the Prevention of Cruelty to Animals prosecuted Thomas E. Carpenter for pouring boiling water on his neighbour's pet chow dog, leaving 'an ugly scar on his back'.[57] Pets were not, therefore, immune from the mistreatment commonly meted out to non-human animals (see Chapter 7), and could easily lose their privileged status. In 2021 an estimated 6.3 million companion animals were relinquished to shelters in the United States of America, of which a significant proportion (920,000) were euthanised.[58]

Affection

So far we have painted quite a bleak picture of pet-keeping, in which human priorities take centre stage. This is only half of the story, however, for pets have also been the objects of intense human affection, loved in life, cared for in sickness and mourned in death. What do historical sources tell us about these more compassionate aspects of pet-keeping? What efforts did pet owners make to protect, nurse and commemorate their animal companions? How far do such exhibitions of affection compensate for the more systemic cruelties inflicted on household pets?

If we study personal sources such as letters, diaries, memoirs and (some) classified advertisements, we certainly find strong evidence that many owners genuinely loved their pets and considered them part of the family. Some penned charming descriptions of their animals, recounting their habits and idiosyncrasies. Some, forced to sell their pets, made a point of seeking good homes for them, apparently putting their welfare above their market value. Others placed advertisements for lost or stolen pets and offered substantial rewards for their safe return. In 1777, for instance, Londoner Lady Austen posted an advertisement for a 'small yellow and white SPANIEL, yellow ears, and three or four yellow marks on his body; is pretty fat, and the hair rather curled, long white hair about his neck and throat, has a short back and short legs and tail, a roundish head, and short nose; has a yellow spot on his forehead, answers to the name of JUBA ... One Guinea Reward'.[59] In 1919 a desperate New Yorker offered a '$100 reward for the return or information leading to [the] recovery of [a] Pekinese dog, lost from [a] Cadillac car near [the] Knickerbocker Hotel, Aug. 14. All information confidential; owner heartbroken'.[60] How far these examples demonstrate true affection may, of course, be questioned. Lady Austen may have been hoping to reclaim valuable property, rather than to retrieve a beloved companion. The missing Pekingese was a pedigree breed, originally from China, and a status symbol in Western countries in the early twentieth century; it was also likely over-pampered (a suggestion reinforced by the fact that it was lost from a Cadillac).[61] Both owners, however, expressed seemingly genuine concern for their pets' well-being and had clearly invested considerable time in training, petting and, in Lady Austen's case, (over) feeding their animal companions, suggesting that more than money was at stake in their recovery.

If attempts to retrieve lost or stolen pets suggest concern for their welfare, so, too, do some owners' efforts to medicate their animals when they were sick. Robert Coquina, Bishop of Durham (1274–1283) fed his two pet monkeys on peeled almonds, served on a silver spoon.[62] Isabella Farnese, the wife of Philip V of Spain (1692–1796), spent 3 reales on an unspecified 'spirit to bathe a cat named Zoquete that had fallen from a slate roof'.[63] One Edwardian bird-lover cured his cockatoo of a 'terribly swollen and inflamed leg' by administering daily massages, while another treated his parrot's 'severe...bronchitis' with a concoction of gruel and warm blackcurrant jam.[64] In 1949, when Australian Danny O'Brien's pet kangaroo 'broke one of its hoppers during the night', he hired a local physician to encase the limb in plaster, aware that the animal, 'a firm favourite with [his] two small sons', would be mourned if it died.[65] Such treatments may not always have been effective; some may even have been counterproductive. The intention behind their administration, seems sincere, however, and reflected a desire to ameliorate suffering and prolong the lives of beloved companions – including quite common animals like cats that could presumably have been cheaply replaced. Some pet owners, indeed, were sufficiently concerned about the well-being of their charges that they requested assistance from fellow owners, seeking information on how best to treat their ailments. In 1787 one Parisian man wrote to the Journal de Paris in 1787 soliciting advice on how to stop his monkey from biting its tail (respondents suggested thrashing it, feeding it an entirely vegetarian diet and/or feeding it snails).[66] A century later Georgina Weldon requested guidance from readers of the RSPCA's *Animal World* magazine on how to care for her pet marmosets, Paddy, Karky and Andy, who, despite being 'kept clean, combed and brushed daily', were suffering from paralysis (she was advised to give her animals 'a warm bath and a dose of castor oil', to line their baskets with 'two light eiderdown cushions' and to provide them with 'a hot water bottle at night; [a] stone ink bottle covered with flannel is the proper thing').[67]

Finally, when the worst happened and pets expired, there is strong evidence that many were mourned, missed and remembered. Letters and diary entries from a variety of periods and places contain loving references to departed animal companions, suggesting that their passing mattered to their owners. Portraits of individual pets were commissioned with increasing frequency from the eighteenth century, in part as a way of immortalising cherished companions, while pet burials have been common across time, from Roman Gaul to nineteenth-century Britain, where the first pet cemetery was established in Hyde Park in 1880.[68] In 1462, for instance, Lodovico II Gonzaga, Marchese of Mantua had his dog Rubino buried in a casket with a tombstone and an epitaph in a spot that he could see from the window of his palace.[69] Not all owners grieved for their pets in this way, of course, and many probably viewed their non-human companions as disposable, or at least replaceable, much like other material goods. Before the opening of a dedicated pet cemetery in Paris in 1899, dogs and cats were frequently sent to a knacker or dumped in streets, posing a significant public health problem.[70] A good proportion of pet owners did (and do), however, perceive their animals as valued individuals, and the elegies written in their honour often bear the hallmarks of a sustained and intimate relationship. In 1876, for example, British dog-owner 'G.R.' penned a touching epitaph for his 'favourite terrier', Snap, recalling the dog's 'winning ways' and 'merry bark' and promising to make 'Snappy's grave ... a cherished spot'.[71]

Pets have thus been the subjects of both dominance and affection, eliciting genuine love and care from their owners, but also experiencing physical cruelty, neglect and abandonment. Individual animals have doubtless been loved by their owners and accorded special privileges. Ultimately, though, human–pet relationships have been primarily for the benefit of, and undertaken at the discretion of, the former, and humans have exercised a constant – if sometimes benign – control over the lives (and deaths) of their companions, tipping the balance in favour of dominance. Studies of pet-keeping in different time periods and cultures (especially non-Western cultures) are beginning to nuance our understanding of pets and their treatment, complicating the dominance–affection binary.

Pets and Other Animals

Before concluding our discussion of pet welfare, we should pose two additional questions, both of which are particularly important when placing pet-keeping within its historical context. First: has pet welfare improved over time, or are the continuities in pet-keeping more striking than the changes? Second: has keeping pets changed our relationship with non-human animals more broadly?

When it comes to pet welfare, we generally like to think that we comprehend and treat out pets better in the twenty-first century than we used to in the past. The actual picture, however, is less clear cut. On the one hand, veterinary treatment for pets has certainly improved, giving most modern pets longer and more pain-free lives than they had in earlier eras. The advent of spaying in the early twentieth century has largely put an end to the drowning of unwanted kittens (at least in the West), while the use of pain relief, effective flea treatments, antibiotics and increasingly sophisticated forms of surgery have all helped to improve pets' quality of life and reduce their suffering. Pet euthanasia has also become much less traumatic than it was in the nineteenth century, when one British guide on pet monkeys advised dispatching sickly simians by hitting them on the back of the skull with an iron bar.[72]

On the other hand, it is arguable that modern pets' lives are often artificially extended for the benefit of their owners rather than the well-being of the animal, that an increasing obsession with breed since the late nineteenth century has led to a growing number of congenital defects, reducing quality of life for many pets (especially dogs), and that the rise of urban pet keeping has imposed ever-increasing limits on animal autonomy, from the denial of adequate exercise and socialisation for dogs to the declawing of cats (especially common in the USA). Abandonment remains a common fate among pets across the globe, many of whom end up in shelters, while shocking cases of abuse still intermittently hit the headlines. The trade in exotic pets, furthermore, which has receded (though by no means disappeared) in Europe and the USA, has intensified in Southeast Asia and the Middle East, with devastating consequences for a wide range of endangered species, from African grey parrots to infant orangutans. We are not, therefore, looking at a triumphalist story of unmitigated progress.

A second area of debate centres on whether owning and caring for pets has contributed to greater care for other types of animals and increased awareness of their suffering. Again, the jury is largely out on this question. On the positive side, it is true that living with a pet can alert humans to the cognitive abilities of other species and their capacity to feel human-like emotions. Writing in 1882, for instance, Alexandra Peckover described the 'different noises' made by her pet monkey, Jemima, and the feelings she believed they conveyed, 'a sweet little chirp, like a bird' indicating that she was 'surprised or frightened', 'a kind of gobble' that she was 'very angry' and 'something like a cooing' that 'she … wants to coddle up close to someone and go to sleep'.[73] This recognition of sentience, made possible by an intimate relationship with a pet, has sometimes been extrapolated to animals outside of the home, sparking greater interest in their welfare; if a pet monkey can feel fear, anger and contentment, then why not a wild monkey in the jungle or a captive monkey in a zoo?

On the negative side, however, it appears that many pet owners have been (and still are) able to compartmentalise their interactions with non-human animals, and to love their pets dearly while continuing to exploit the unseen, depersonalised beasts outside the confines of their homes. Many people who love their dogs and cats have simultaneously gone hunting, worn fur and eaten meat, perhaps without even thinking about the anonymous animals killed in the process. A poem published in 1890, for example, condemned the hypocrisy of a fictional fur-wearing duchess, who happily 'rides in skins of seals', allegedly stripped from the animals while still alive, but 'Would weep, and vow it was a shame, / If any dared her pug to hurt'.[74] Pets themselves, furthermore, have become consumers in their own right and often contribute directly or indirectly to the suffering of other animals, whether through predation on wild birds and mammals (in the case of cats) or the consumption of mass-produced pet-foods, which, at different points in the twentieth century, have contained battery-farmed poultry, whale meat and fishmeal from overexploited fish-stocks.[75] Once again, therefore, we are presented with a contradictory picture, in which the ethical status of pets remains unclear.

Cats: Goddesses, Familiars and Favourites

To explore some of these trends in more detail, the chapter concludes with a case study of the cat (*Felis catus*): the world's second most popular pet in the early twenty-first century (just behind dogs). Originally domesticated around 10,000 years ago, cats have enjoyed a love–hate relationship with humans over the centuries, being worshipped as gods and reviled as devils. An analysis of their fluctuating popularity reveals wider shifts in human attitudes towards other animals and highlights the pros and cons of life as an animal companion.

Cats were first domesticated in the Near East, in an area known as the Fertile Crescent. They are descendants of the African wild cat (*Felis sylvestris lybica*). Attracted to human settlements by the presence of mice and rats in granaries, it is thought that cats largely domesticated themselves through a gradual process of natural selection. Bolder and tamer cats would approach humans more closely and prey on vermin, winning human affection. More timid or aggressive felines would be killed by humans, or outcompeted by their tamer, better-fed counterparts. The oldest known example of a domesticated cat was found at Shillourokambos in Cyprus in a grave dating to 7,500BC.[76]

Cats were valued highly by the ancient Egyptians, who admired their grace, fecundity and swiftness. Surviving friezes show cats sitting alongside their human owners eating fish and engaged in hunting wildfowl. There is also evidence that members of the Egyptian elite dressed their cats in gold and allowed them to eat from their plates. When a family cat died, its human owners would shave their eyebrows as a mark of respect and mummify their former companion so that it might enter the afterlife. Killing a cat was punishable by death; Diodorus Siculus saw a mob of Egyptians demand, and apparently secure, the death of a man connected with the Roman embassy in 59BC after he accidentally killed a cat. While domestic cats were venerated, however, high demand for dead cats by religious cults meant that thousands of the animals were killed in ancient Egypt for use as votive offerings – the only context in which cat slaughter was permitted. Analysis of the mummified remains of these animals suggests that they were mostly young animals, bred for the purpose and killed by having their necks broken.[77]

From Egypt, domesticated cats were likely imported to India, and, from there to China, where, by the eleventh century, cat fanciers had begun to develop a specialist vocabulary to describe cats of different colours and to breed the first long-haired domestic cats (called lion cats, or *shimao*).[78] In Europe, meanwhile, cats gradually became a fixture in medieval and early modern households, leaving traces of their presence in buildings and contemporary records. An account book for Cuxham manor in Oxfordshire includes an entry in 1293–1294 for cheese bought for a cat.[79] The seventeenth-century French statesman Cardinal Richelieu built special quarters for his cats in his palace and provided for their upkeep in his will.[80] Unfortunately not all domestic cats were as lucky as Richelieu's pampered pets, and most early modern felines continued to be viewed primarily as working animals. Violence towards cats was also common before 1800, some of it spurred by a superstitious belief in witchcraft. In 1638, for example, a man named William Smith roasted a live cat on a spit in Ely Cathedral.[81] In 1730 a group of printer's apprentices in Paris, tired of being kept awake at night by the sound of cats howling, rounded up the animals, subjected them to a mock trial and executed them by hanging.[82]

In the nineteenth century, cats enjoyed a resurgence of popularity in the West and began to occupy their modern status as pets. Appreciated by the Victorians for their cleanliness, felines were increasingly perceived as household companions, and were treated with greater consideration. Exotic cats like Persians, Siamese and Abyssinians started to be imported into Europe from the 1870s and 80s and were exhibited in cat shows (the first of which took place in 1871). Less distinguished moggies also elicited affection, appearing with growing regularity in portraits and photographs, inspiring touching tributes from their owners when they died and being ministered to by the latter when they were sick or injured. In 1911 when

Maude Burnett's tabby cat was 'frightfully bitten in the hind-leg by a fox-terrier', she kept 'a loose bandage constantly dipped in Condy's fluid wrapped round the leg' and fed him 'on warm milk' (Figure 3.4).[83] As the century progressed, animal welfare organisations also started to encompass domestic cats within their remit, prosecuting acts of suspected cruelty, setting up homes for neglected and abandoned cats and urging cat owners not to leave their animals to fend for themselves when they went on vacation. While the need for such campaigns underlined the abuse still directed at felines in this period, the strong reactions generated by acts of mistreatment suggested a new respect for domestic cats, and a sense that subjecting them to cruelty was wrong. In 1893, for instance, a judge in Washington DC fined a young man named Ernest Brown $5 for picking up 'a confiding and inoffensive tabby cat by the tail, and after whirling it around his head let[ting] it fly by its acquired momentum into a tree top'.[84] In 1923 magistrates in Birmingham sentenced William Henry Ratchford and Ernest Bailey to 'a month's hard labour' for tying 'a salmon tin so tightly to [a] cat's tail that a swelling was caused'.[85]

Better understood today than they once were, pet cats are increasingly coddled, benefiting from improved veterinary medicine and nutrition. They are also, however, subject to more intensive forms of bodily control, from defleaing and spaying to declawing, and are often compelled to spend their entire lives indoors. Not as easy to breed as dogs, cats have suffered less from bodily manipulation at the hands of humans and have diverged less dramatically

Figure 3.4 'Pussy's Bad Leg', *The Animals' Friend*, 1911, p. 72.
Source: photograph by author.

from their original form. Several new breeds have, nonetheless, emerged in the twentieth century, some of which suffer from genetic defects. The Sphynx, for instance (created in the 1960s), is completely hairless, and therefore highly susceptible to excessive cold or heat. The Californian ragdoll (created in the 1970s) goes limp when picked up by humans, so lacks the normal feline mechanisms for escaping abuse. At the other end of the spectrum, large numbers of stray cats live on the streets, enduring neglect, starvation, disease and violence. Outdoor and feral cats, meanwhile, kill large numbers of birds and rodents, raising ethical questions about the relative rights of companion animals and their wild counterparts; as one early twentieth-century North American bird lover remarked, 'oceans of cream and miles of blue ribbon have not subdued Pussy's instincts for the chase, nor destroyed her skill as a hunter'.[86]

The changing fortunes of the domestic cat illustrate the complex and fluctuating status of companion animals. Valued, by turns, for their hunting capabilities, their aesthetic qualities and their (qualified) friendship, cats have been worshipped, loved and pampered, enjoying the privilege of pet status. Perceived, on the other hand, as 'selfish, treacherous and absorbed in their own comfort' (to cite one Victorian commentator), domesticated felines have simultaneously been victims of cruelty and persecution, whether at the hands of sadistic youths or religious zealots.[87] They have also suffered from the attentions of ignorant or over-bearing owners, who have curtailed their freedom, manipulated their bodies and fed them inappropriate diets; in 1925, for instance, the Massachusetts Society for the Prevention of Cruelty to Animals prosecuted a Professor at Boston University for feeding his kitten on 'sour milk, cooked bacon and peanut butter sandwiches'.[88] The experience of domestic cats thus highlights the complex dynamics of pet-keeping and its capacity to privilege and control non-human animals. Adored, cossetted, anthropomorphised, denatured and discarded, cats have enjoyed the benefits of living closely with humans but also endured the consequences of human superstition, interference and neglect.

Conclusion

Pets have been present in almost every human society. Once largely the preserve of the elite, they were adopted by the urban bourgeoisie in nineteenth-century Europe, and later by the working classes. In the twentieth century, Western-style pet keeping spread across the globe, reshaping human–animal relations in Japan, China, India and Nigeria.

Pets offer unusually rich subject material for historians thanks to the abundance of source material they leave behind. Depicted in paintings, captured on camera, referenced in letters, diaries and memoirs and advertised in newspapers, pets leave a deeper paw-print in the historical record than most other animals, allowing scholars to more fully reconstruct their biographies.

In so doing, historians have documented the affection many owners felt towards their animal companions and the grief experienced when they died. They have also revealed some of the darker aspects of pet-keeping in past societies, from physical abuse to misguided feeding practices; one pet wallaby in 1920s Western Australia was regularly treated to 'a hot dinner – Yorkshire pudding and cooked vegetables and gravy; also a little fish' (a diet which may have contributed to paralysis in the marsupial's hindlegs).[89] Though atypical of the wider animal experience, pets reflect the close bond that exists between some humans and their animals and the potential for intimate cross-species connections. As both commodities and consumers, animal companions provide a fascinating window into the complex and contradictory nature of human–animal relationships, offering unique insights into interspecies interactions.

Notes

1 Keith Thomas, *Man and the Natural World: Changing Attitudes in England, 1500–1800* (London: Penguin, 1983), pp. 6–7.
2 Katherine Grier, *Pets in America: A History* (Chapel Hill: University of North Carolina Press, 2006), pp. 231–270.
3 Kathleen Walker-Meikle, *Medieval Pets* (Woodbridge: The Boydell Press, 2012), p. 12.
4 Pliny the Elder, *Natural History*, ed. J. Henderson and trans. W.H.S. Jones (Cambridge, MA: Harvard University Press, 1975), Book 10, ch.58.
5 Marta Osypinska, 'Pet Cats at the Early Roman Red Sea port of Berenike, Egypt', *Antiquity* 90: 354, e5 (2016), pp. 1–5.
6 Sarah Cockram, 'Sleeve Cat and Lap Dog: Affection, Aesthetics and Proximity to Companion Animals in Renaissance Mantua', in Sarah Cockram and Andrew Wells (eds), *Interspecies Interactions: Animals and Humans from the Middle Ages to Modernity* (London: Routledge, 2018), pp. 34–65.
7 Letter from Catherine the Great to Friedrich Melchoir Grimm, 23 November 1785, in Andrew Kahn and Kelsey Rubin-Detlev (eds), *Catherine the Great: Selected Letters* (Oxford: Oxford University Press, 2018), p. 251.
8 Grier, *Pets in America*, pp. 231–270; 'A Jaunt to Jamrach's', *The Era*, 13 September 1884.
9 Kathleen Kete, *The Beast in the Boudoir: Petkeeping in Nineteenth-Century Paris* (Berkeley: University of California Press, 1994), pp. 76–96.
10 'Daft on Dogs' Clothes', *Chicago Daily Tribune*, 6 September 1896.
11 Neil Pemberton, Julie-Marie Strange, and Michael Worboys, *The Invention of the Modern Dog: Breed and Blood in Victorian Britain* (Baltimore: Johns Hopkins University Press, 2018), pp. 90–114.
12 Aaron Skabelund, *Empire of Dogs: Canines, Japan and the Making of the Modern Imperial World* (Ithaca: Cornell University Press, 2011), pp. 53–86.
13 Peter J. Li, *Animal Welfare in China* (Sydney: University of Sydney Press, 2021), pp. 7–8.
14 Julie Lockwood et al., 'When pets become pests: the role of the exotic pet trade in producing invasive vertebrate animals', *Front Ecol Environ* 17:6 (2019), pp. 323–330.

15 'Python Wars: The Snake Epidemic Eating Away at Florida', *The Guardian*, 22 August 2019.
16 Grier, *Pets in America*, pp. 39–42.
17 Sarah Amato, *Beastly Possessions: Animals in Victorian Consumer Culture* (Toronto: University of Toronto Press, 2015), p. 55.
18 Helen Cowie, *Victims of Fashion: Animal Commodities in Victorian Britain* (Cambridge: Cambridge University Press, 2022), pp. 196–233.
19 'Country House', *The Bazaar*, 9 March 1900, p. 858.
20 John Rippingham, *Natural History According to the Linnaean System, Explained by Familiar Dialogues in Visits to the London Museum* (London: N. Hailes, 1817), vol. IV, pp. 14–15.
21 'My Pet Wallaby', *The Age*, 5 July 1935.
22 An Unusual Family Pet in a Pennsylvanian Home: Teddy the Giant Anteater', *The Illustrated London News*, 13 March 1954.
23 'Orders Safety Belt for his Pet Alligator', *Chicago Tribune*, 9 April 1963.
24 Chris Pearson, *Dogopolis: How Dogs and Humans Made Modern New York, London and Paris* (Chicago: University of Chicago Press, 2021), pp. 47–82; Saheed Aderinto, *Animality and Colonial Subjecthood in Africa* (Athens: Ohio University Press, 2022), pp. 145–174.
25 C. Cumberland, *The Guinea Pig, or Domestic Cavy, for Food, Fur and Fancy* (London: L. Upcott Gill, 1886), pp. 36–43.
26 'A Pet Kangaroo', *Western Mail*, 20 August 1936.
27 Erick Laurent, 'Children, "Insects" and Play in Japan', in Anthony Podberscek, Elizabeth Paul and James Serpell (eds), *Companion Animals and Us: Exploring the Relationships between People and Pets* (Cambridge: Cambridge University Press, 2000), pp. 61–89.
28 Marcy Norton, *The Tame and the Wild: People and Animals after 1492* (Cambridge, MA: Harvard University Press, 2024), pp. 134–135.
29 Henry Walter Bates, *The Naturalist on the River Amazons* (London: John Murray, 1864), p. 394.
30 Norton, *The Tame and the Wild*, p. 148.
31 Walker-Meikle, *Medieval Pets*, p. 7.
32 Rippingham, *Natural History*, Vol. IV, p. 55.
33 Ingrid Tague, *Animal Companions: Pets and Social Change in Eighteenth-Century Britain* (Philadelphia: Penn State University Press, 2015), pp. 91–137.
34 'Extraordinary Collection of Cats: A Lady Artist and her Menagerie', *Illustrated Police News*, 1 June 1867.
35 'Dogs in Prams: Taiwan's Falling Birth-rate Sees Pets Outnumbering Children', *The Guardian*, 6 November 2020. Japan is experiencing a similar trend; see Skabelund, *Empire of Dogs*, p. 189.
36 Juliana Schiesari, '"Bitches and Queens": Pets and Perversion at the Court of France's Henri III', in Erica Fudge (ed.), *Renaissance Beasts: Of Animals, Humans and Other Wonderful Creatures* (Urbana: University of Illinois Press, 2004), pp. 37–49.
37 'Tuesday's Post', *Ipswich Journal*, 10 May 1828.
38 'The Mirror of Fashion' *The Morning Chronicle* 15 July 1828.
39 'Correspondence', *Liverpool Mercury*, 7 August 1829.
40 Philip Howell, *At Home and Astray: The Domestic Dog in Victorian Britain* (Charlottesville: University of Virginia Press, 2015), pp. 125–149.
41 Zeb Tortorici, '"In the Name of the Father and the Mother of All Dogs": Canine Baptisms, Weddings and Funerals in Bourbon Mexico', in Martha Few and Zeb

Tortorici (eds), *Centering Animals in Latin American History* (Durham, NC: Duke University Press, 2013), p. 102.

42 Louise Robbins, *Elephant Slaves and Pampered Parrots: Exotic Animals in Eighteenth-Century Paris* (Baltimore: John Hopkins University Press, 2002), p. 198.

43 'Cruelty to Parrots', *The Animals' Friend*, 1907, p. 109.

44 Yi-Fu Tuan, *Dominance and Affection: The Making of Pets* (New Haven: Yale University Press, 1984), pp. 88–161.

45 'My Monkeys', *Morning Post*, 26 December 1867.

46 C.W. Gedney, 'The Grey Parrot', *The Bazaar*, 3 February 1877, p. 76.

47 'Cage Birds', *The Bazaar*, 27 September 1893, p. 801.

48 On the twentieth- and twenty-first-century trade in exotic pets, see Rosemary-Claire Collard, *Animal Traffic: Lively Capital in the Global Exotic Pet Trade* (Durham: Duke University Press, 2020).

49 'Super Furry Primate is Under Threat from Japanese Pet Trade', *The Times*, 5 February 2016.

50 Jane Hamlett and Julie-Marie Strange, *Pet Revolution: Animals and the Making of Modern British Life* (London: Reaktion, 2023), pp. 19–47.

51 'Copy-Cats: Pet-Cloning in China', *The Guardian*, 6 September 2019.

52 'Parrots in Captivity', *The Animal World*, March 1893, p. 47.

53 Carlos Gómez Centurión-Jiménez, 'Chamber Animals at the Spanish Court during the Eighteenth Century', *The Court Historian* 16:1 (2011), pp. 54–55.

54 'Dinah', *The Animal World*, April 1894, pp. 172–13.

55 'Provide for your Cats before taking Holiday', *The Animal World*, March 1897, p. 128; Grier, *Pets in America*, pp. 81–82.

56 Annie Martin, *Home Life on an Ostrich Farm* (London: George Philip and Son, 1890), pp. 243–244.

57 'Scalded Dog Takes the Stand', *Washington Post*, 27 June 1936.

58 ASPCA, 'Pet Statistics' (undated), retrieved from www.aspca.org/helping-people-pets/shelter-intake-and-surrender/pet-statistics.

59 'A Dog Lost', *Morning Chronicle and London Advertiser*, February 10, 1777.

60 'Lost and Found – Cats and Dogs', *New York Times*, 19 August 1919.

61 Sarah Cheang, 'Women, Pets and Imperialism: The British Pekingese Dog and Nostalgia for Old China', *Journal of British Studies* 45 (2006), pp. 359–387.

62 Walker-Meikle, *Medieval Pets*, p. 34.

63 Gómez Centurión-Jiménez, 'Chamber Animals at the Spanish Court', p. 56.

64 'A Cure by Massage', *The Animal World*, December 1905, p. 191; 'Treatment of Parrots Affected with Bronchitis', *The Animal World*, October 1902, p.159.

65 'Hopper of Pet Kangaroo Placed in Plaster', *Quorn Mercury*, 3 February 1949.

66 Robbins, *Elephant Slaves and Pampered Parrots*, p. 136.

67 'To Monkey Fanciers', *The Animal World*, September 1886, p. 158; 'Bearing Reins, Monkeys and Dogs', *The Animal World*, November 1886, p. 174.

68 Teresa Mangum, 'Animal Angst: Victorians Memorialize their Pets', in Deborah Denenholz Morse and Martin Danahay, *Victorian Animal Dreams* (Farnham: Ashgate, 2007), p. 24.

69 Cockram, 'Sleeve Cat and Lap Dog', p. 49.

70 Robbins, *Elephant Slaves and Pampered Parrots*, p. 90.

71 'Epitaph on a Favourite Terrier', *The Animal World*, January 1876, p. 11.

72 Arthur Patterson, *Notes on Pet Monkeys and How to Manage Them* (London: L. Upcott Gill, 1888), p. 101.

73 'Jemima', *The Animal World*, May 1882, p. 75.

74 'A Protest in Rhyme Against Seal Cruelty', *The Animal World*, March 1890, p. 39.
75 On the ethics of feeding meat to companion animals, see Josh Milburn, *Just Fodder: The Ethics of Feeding Animals* (Montreal: McGill-Queen's University Press, 2022), pp. 50–81.
76 Brian Fagan, *The Intimate Bond: How Animals Shaped Human History* (London: Bloomsbury 2015), p. 67.
77 Juliet Clutton-Brock, *Animals as Domesticates: A World View Through History* (East Lansing: Michigan State University Press, 2015), pp. 59–60.
78 Timothy Barrett and Mark Strange, 'Walking by Itself: The Singular History of the Chinese Cat', in Roel Sterckx, Martina Siebert and Dagmar Schäfer (eds), *Animals Through Chinese History: Earliest Times to 1911* (Cambridge: Cambridge University Press, 2019), pp. 84–98.
79 Walker-Meikle, *Medieval Pets*, p. 10.
80 Fagan, *The Intimate Bond*, p. 223.
81 Thomas, *Man and the Natural World*, p. 109.
82 'Workers Revolt: The Great Cat Massacre of the Rue Saint Severin', in Robert Darnton, *The Great Cat Massacre: And Other Episodes in French Cultural History* (London: Allen Lane, 1984), pp. 75–104.
83 'Pussy's Bad Leg', *The Animals' Friend*, 1911, p. 72.
84 'Fined for Cruelty to Cats', *Washington Post*, 7 July 1893.
85 'Cruelty to a Cat', *Devon and Exeter Gazette*, 21 August 1923.
86 'The Cat Question', *Bird-Lore*, April 1902, p. 70; Peter P. Marra and Chris Santella, *Cat Wars: The Devastating Consequences of a Cuddly Killer* (Princeton: Princeton University Press, 2016).
87 'Are Cats Selfish?', *The Animal World*, January 1884, p. 30.
88 'Judge Rules on Cat's Food', *New York Times*, 23 June 1925.
89 'Pet Wallaby Paralysed', *Western Mail*, 19 December 1929.

Further Reading

Barrett, Timothy, and Mark Strange, 'Walking by Itself: The Singular History of the Chinese Cat', in Roel Sterckx, Martina Siebert and Dagmar Schäfer (eds), *Animals Through Chinese History: Earliest Times to 1911* (Cambridge: Cambridge University Press, 2019), pp. 84–98
Cheang, Sarah, 'Women, Pets and Imperialism: The British Pekingese Dog and Nostalgia for Old China', *Journal of British Studies* 45 (2006), pp. 359–387
Cockram, Sarah, 'Sleeve Cat and Lap Dog: Affection, Aesthetics and Proximity to Companion Animals in Renaissance Mantua', in Sarah Cockram and Andrew Wells (eds), *Interspecies Interactions: Animals and Humans from the Middle Ages to Modernity* (London: Routledge, 2018), pp. 34–65
Collard, Rosemary-Claire, *Animal Traffic: Lively Capital in the Global Exotic Pet Trade* (Durham: Duke University Press, 2020)
Cowie, Helen, 'Monkey Business', in Helen Cowie, *Victims of Fashion: Animal Commodities in Victorian Britain* (Cambridge: Cambridge University Press, 2022), pp. 196–233
Gómez Centurión-Jiménez, Carlos, 'Chamber Animals at the Spanish Court during the Eighteenth Century', *The Court Historian* 16: 1 (2011), pp. 43–65
Grier, Katherine, *Pets in America: A History* (Chapel Hill: University of North Carolina Press, 2006)

Hamlett, Jane, and Julie-Marie Strange, *Pet Revolution: Animals and the Making of Modern British Life* (London: Reaktion, 2023)

Howell, Philip, *At Home and Astray: The Domestic Dog in Victorian Britain* (Charlottesville: University of Virginia Press, 2015)

Kete, Kathleen, *The Beast in the Boudoir: Petkeeping in Nineteenth-Century Paris* (Berkeley: University of California Press 1994)

Norton, Marcy, 'Taming Strangers', in Marcy Norton (ed), *The Tame and the Wild: People and Animals after 1492* (Cambridge, MA: Harvard University Press, 2024), pp.130–149

Pearson, Chris, *Dogopolis: How Dogs and Humans Made Modern New York, London and Paris* (Chicago: University of Chicago Press, 2021)

Pemberton, Neil, Julie-Marie Strange and Michael Worboys, *The Invention of the Modern Dog: Breed and Blood in Victorian Britain* (Baltimore: Johns Hopkins University Press, 2018)

Podberscek, Anthony, Elizabeth Paul and James Serpell, *Companion Animals and Us: Exploring the Relationships between People and Pets* (Cambridge: Cambridge University Press, 2000).

Schiesari, Juliana, '"Bitches and Queens": Pets and Perversion at the Court of France's Henri III', in Erica Fudge (ed.), *Renaissance Beasts: Of Animals, Humans and Other Wonderful Creatures* (Urbana: University of Illinois Press, 2004), pp. 37–49

Skabelund, Aaron, *Empire of Dogs: Canines, Japan and the Making of the Modern Imperial World* (Ithaca: Cornell University Press, 2011)

Tague, Ingrid, *Animal Companions: Pets and Social Change in Eighteenth-Century Britain* (Philadelphia: Penn State University Press, 2015)

Thomas, Keith, 'Domestic Companions' and 'Privileged Species', in Keith Thomas, *Man and the Natural World: Changing Attitudes in England, 1500–1800* (London: Penguin, 1983), pp. 91–101

Tortorici, Zeb, '"In the Name of the Father and the Mother of All Dogs": Canine Baptisms, Weddings and Funerals in Bourbon Mexico', in Martha Few and Zeb Tortorici (eds), *Centering Animals in Latin American History* (Durham, NC: Duke University Press, 2013), p. 93–122

Tuan, Yi-Fu, *Dominance and Affection: The Making of Pets* (New Haven: Yale University Press, 1984)

Walker-Meikle, Kathleen, *Medieval Pets* (Woodbridge: The Boydell Press, 2012)

4 Exhibition

Exotic animals have long fascinated humans. For millennia they have been collected and exhibited in menageries and zoological gardens. They have been paraded, studied and trained to perform, functioning by turns as symbols of monarchical or imperial power, items of trade, subjects for scientific enquiry and sources of entertainment.

The compulsion to collect exotic animals appears to have been almost universal across time and space. The Pharoah Hatshepsut (c.1504–1457BC) kept a menagerie at Thebes which included elephants and leopards from India and a giraffe from Somalia.[1] The Chinese emperor Wen Wang (1046–256BC) constructed a 375-acre park housing birds, deer and fish. The kings of Assyria and Persia kept an array of animals in the first century BC for hunting, sacrifice and exhibition, while the Romans held large numbers of exotic animals captive for use in bloody gladiatorial combats.[2] According to the Spanish conquistador Bernal Díaz, the Aztec emperor Moctezuma II had a large menagerie in Tenochtitlán containing eagles, quetzals, 'tigers [jaguars] and two kinds of lions [pumas], and animals something like wolves and foxes', the latter allegedly fed on the bodies of sacrifice victims.[3] Since the early modern era menageries and zoos have functioned by turns as emblems of empire, commercial enterprises and, recently, centres of conservation, their stock reflecting shifting trade patterns and transoceanic connections. Some captive animals have also become global stars, travelling between continents and attaining transnational celebrity. The famous elephant Jumbo was captured in Abyssinia (modern-day Ethiopia) in 1861, exhibited in the zoological gardens of Paris and London and toured across the eastern states of the USA by the showman Phineas Taylor Barnum until his death in Ontario, Canada, in 1886.

Until relatively recently, most zoo histories were institutional biographies, often written by zoo professionals. These works charted the development of specific zoological collections and tended to tell a story of gradual progress over time, with larger enclosures and improved architecture leading to better conditions for captive animals. Over the last few decades, more critical

DOI: 10.4324/9781003181996-5

studies of zoos have emerged, challenging the 'Whiggish' narrative of these earlier histories. Some of these works are explicitly anti-zoo in their approach, influenced by the growing animal rights movement. Others are less overtly political, posing new questions about zoological institutions and situating them within broader historical fields. Historians have, for instance, examined the relationship between zoos and monarchical/imperial power, assessed the role of zoos as venues for scientific research, studied the role of zoos as places of entertainment and explored the connections between zoos and civic identity. The zoo has thus become a prism through which to study wider aspects of human culture, attracting the attention of historians of empire, historians of leisure and historians of science.

Power

A recurrent theme in the history of zoos and menageries has been the relationship between exotic animals and different forms of power. Historians have seen zoos as symbolsof man's control over the natural world, on the one hand, and, on the other, of human control over other humans, be they monarchical subjects or colonial peoples. Though the form and content of menageries has changed over time, these functions have remained to the fore, mediating interactions with exotic animals. As zoo critic Randy Malamud observes:

> Founding and operating a zoo involves both real and metaphorical appropriative control of the earth: of nature, land and habitat, and of animals taken from natural habitats, subjugated and recontextualized in a way that upholds the captors' self-serving ideologies.[4]

In the medieval and early modern periods, exotic animals often functioned as diplomatic currency. Embodying the novel and the rare, they were exchanged between monarchs as symbols of deference or allegiance, serving as valued gifts alongside spices, perfumes and precious metals.[5] In 1252, for instance, King Haakon of Norway gave Henry III of England a young polar bear, who was lodged in the Tower of London and provided with a 'muzzle and chain' to allow him to fish and wash himself in the River Thames.[6] In 1487 the Sultan of Cairo, al-Ashraf Quaitbay, presented the Florentine noble Lorenzo de Medici with a young giraffe, who was 'carried through various parts of Tuscany and exhibited in several convents' (Figure 4.1).[7] In 1678 the Portuguese envoy Bento Pereira presented the Emperor K'ang-hsi of China with an African lion, bringing the animal by sea from Mozambique and conveying it overland from Macao to Beijing.[8] The association of these animals with distant and little-known places contributed heavily to their attraction, adding to their exoticism and allure. The sheer difficulty of transporting

Figure 4.1 Giorgio Vasari, *Lorenzo the Magnificent Receives the Tribute of the Ambassadors*, Palazzo Vecchio, Florence, 1556–1558.

Source: Album / Alamy Stock Photo.

them alive across deserts and oceans further enhanced their appeal, under-lining the power and resources of their owners. An elephant sent to Charles III of Spain by the Governor of the Philippines in 1773 required a daily ration of '85 quartillos of water, 24 pounds of rice, 6 pounds of sugar, 2 rations of wine, 2 and a half ordinary rations of bread and 4 servings of bananas' during its six-month voyage to Cádiz, and was medicated three times a month with 'thirty different spices'.[9]

If the ability to acquire coveted species attested to the power of monarchs and nobles, their potency was further underlined by the settings in which they showcased their living possessions. As evidence of their control over the natural world – and, by extension, their subjects – some princes used their exotic animals in grandiose processions or deployed them in bloody animal baits for the entertainment of their courts. In the early seventeenth century, the Ottoman Emperor Murad IV choreographed a parade of 'Ten lions, five leopards, twelve tigers and a group of hyenas, foxes, wolves and jackals' through the centre of Istanbul to commemorate a recent military cam-paign.[10] In 1631 Philip IV of Spain staged a fight between a lion and a bull in a specially constructed amphitheatre for the entertainment of the royal court.[11] As the seventeenth century wore on, such visceral displays of power became less common, and more peaceful demonstrations of authority took centre stage. Well-stocked menageries came to form part of the theatre of power for ambitious European monarchs, with novel display techniques allowing visitors to appreciate their control over the natural world. Louis XIV's large menagerie at Versailles led the way in architectural innovation, its semi-circular layout permitting the viewer to survey all the animals at once and to bask in the reflected glory of the Sun King.[12]

The early nineteenth century witnessed a shift in focus from monarchical to national power. This shift began in 1793, when French revolutionaries transferred the surviving animals from the menagerie at Versailles to a new national menagerie in the Parisian Jardin des Plantes. It was consolidated in 1828 with the establishment of the Gardens of the Zoological Society of London (the future London Zoo), and the creation of municipal zoos in cit-ies such as Dublin (1831), Bristol (1835), Antwerp (1843), Berlin (1844), Marseilles (1854), Budapest (1865), New York (1864), Philadelphia (1874), Poznán (1874), Calcutta (1875), Tokyo (1882), Buenos Aires (1888) and Barcelona (1892). Accessible to the general public (though not without restrictions), zoos acted as symbols of national potency and commercial strength and often competed with one another for the most novel and cov-eted animals, taking advantage of steam shipping and railways to import them from ever-longer distances; a Galapagan tortoise exhibited in London Zoo in 1890 was shipped to Britain from Sydney on 'the Peninsular and Oriental mail steamship *Oceana*' and transported from Plymouth to London by rail, stopping en route at Bristol for 'a fresh supply of footwarmers, which had been telegraphed for' in advance.[13] Recent scholarship has also emphasised the link between zoos and civic pride, noting their direct interactions with local communities. In 1890 the *Atlanta Constitution* solicited donations to purchase an elephant for the city's Grant Park Zoo, publishing the portrait of any child who raised more than $10 and charging 5 cents for citizens to vote on the animal's name (the pachyderm, which arrived in August 1890, was eventually christened Clio).[14]

At the same time as zoological gardens were appearing across Europe and North America, the first travelling menageries and circuses also started to tour both continents, transporting exotic animals by road and, later, by rail.[15] Initially focused, like zoos, on the simple display of wild beasts, circuses soon began to place more emphasis on their animal performances, training lions, bears and elephants to execute outlandish – and often dangerous – tricks. In 1885 an elephant named John L. Sullivan in Forepaugh's Circus was 'taught to box with his trainer', Ephraim Thompson, jabbing at his human opponent with a boxing glove attached to the end of his trunk.[16] In 1901 a trained sea lion named Frisco performed in a travelling circus with Captain James Woodward, balancing a ball on his nose and beating a drum with a stick fastened to his flipper.[17] Commercial entities rather than public institutions, menageries prioritised entertainment over education, emphasising the ferocity of their wild inmates and tempting visitors with sensational feats and cheap thrills. Thanks to their extensive itineraries, these shows were, nonetheless, important sources of basic information about exotic animals, providing many people with their first glimpse of exotic beasts. Writing in 1858, the *Bristol Mercury* remarked that 'even in these days, although Bristol and a handful of the leading towns can boast of their Zoological Gardens, there are scores of communities … who would never see a lion, an elephant or a rhinoceros if these menageries were driven off the road'.[18]

Though different in emphasis, both zoos and menageries have been perceived by historians as emblems of imperial power, and key sites for transmitting the achievements of empire to domestic populations. By gathering hundreds of different species in a single locale, zoos made empire tangible to metropolitan audiences, who could live the empire vicariously through a visit to the menagerie. By drawing on a range of imperial personnel to obtain exotic animals, zoos also reinforced their colonial connections, enabling spectators to visualise distant colonial settings and to associate certain animals with specific agents of empire or theatres of imperial action. A Bactrian camel exhibited at Dublin Zoo in 1856, for instance, was reportedly 'taken in the flank search after the battle of Alma' in the Crimean War and presented to the Royal Society of Ireland by Doctor William Carte, having served 'with the British army until the close of the war'.[19] As overseas empires expanded, European zoos were increasingly able to exploit their colonial connections, making use of their countries' varied colonial portfolios to access different exotic fauna. Antwerp Zoological Garden specialised in species from the Congo Free State, such as gorillas and okapis. France established a zoological garden in Antananarivo in 1932, largely so that it could act as a depot for forwarding Madagascar's unique fauna to Paris.[20] From the mid-nineteenth century, professional animal dealers also emerged in port cities such as Liverpool, Marseilles and Hamburg, drawing on a global network of agents in Africa, Asia and the Americas to import giraffes, elephants and rhinoceroses on an industrial scale. Carl Hagenbeck, the most famous animal dealer of his era, employed agents in Siberia, Central America, Thibet, the Sudan and the Philippines and counted Victor Emmanuel II of Italy, the Emperor of Austria and the Mikado of Japan among his customers.[21]

While empire has dominated the history of zoological attractions, some more recent historiography has challenged this focus, suggesting that the relationship between zoos and imperial culture needs nuancing. On the one hand, there is a question mark over the contents of zoological exhibitions, which did not necessarily emanate purely from colonial possessions. London Zoo, for instance, exhibited many South American animals – most notably a giant anteater in 1853 – yet South America was never formally colonised by Britain. The Bogd Khan of Mongolia purchased an elephant from a Russian circus in 1913, which he kept in the winter palace in Ulaanbaatar and fed on steamed dumplings and red plums.[22] What was being exhibited in these cases was therefore a fascination with a generic 'exotic', rather than an explicit evocation of empire.

On the other hand, as Bob Mullan and Garry Marvin have pointed out, zoos were not confined to colonizing countries, but also emerged in the cities of former colonies, such as Buenos Aires, and in European nations that never possessed an overseas empire, such as Poland and Hungary.[23] Some of these institutions focused on the acclimatisation of non-native animals. Others imitated their imperialist rivals by displaying a classic array of charismatic megafauna (especially elephants), and engaging in what historian Marianna Szczygielska describes as 'second hand orientalism'.[24] Yet others concentrated primarily on showcasing native wildlife, which they were well-placed to collect; the National Zoo in Washington DC focused on the conservation of native North American species, such as the bison, currently under threat in the American West. Zoos thus represented local, regional, commercial or national power as well as imperial power and acted as markers of transcontinental trade, internal colonisation (in the case of the USA) and cultural sophistication.

The days of formal empire are now over, and zoos no longer function as overt symbols of monarchical, national or imperial power. This does not mean, however, that power-relations are absent from the modern zoo. On the contrary, they are central to its management practices, shaping both how animals are perceived by the visiting public and how they exist on a day-to-day basis.

First, when it comes to sourcing exotic animals, many postcolonial zoos have continued to extract animals from their former colonial possessions, taking advantage of pre-existing relationships. Antwerp Zoo, for instance, continued to receive animals from the Congo (then Zaire) during the dictatorship of President Mobutu, while the Zoological Society of London (ZSL) received twenty white rhinos from the Umfolozi reserve in Natal in 1970 for exhibition at Whipsnade Park Zoo (an offshoot of London Zoo founded in 1931).[25] China, meanwhile, has engaged in 'panda diplomacy' since the 1940s, sending specimens of the cuddly black and white bears to zoos overseas as a form of soft power.[26] As in the nineteenth century, new forms of transport – in this case air travel – have facilitated the movement of movement of exotic animals between continents, serving, once again, as a sign of technological progress; in 1955, seven Emperor and four Adelie penguins travelled by plane from Buenos Aires to Miami en route to Washington's National Zoological Park, feasting on 'generous quantities of fresh fish' and being 'doused constantly with cold water' to keep them cool.[27]

Modern zoos also continue to exercise a form of stewardship over their animals which Irus Braverman describes as 'pastoral power'. This stewardship extends beyond zoo inmates to their counterparts in the wild and forms part of a wider drive to protect and conserve endangered species. It entails close control and surveillance of zoo animals and involves 'seven interrelated technologies of animal governance; naturalizing, classifying, seeing, naming, registering, regulating and ... collectively reproducing zoo animals'.[28] While zoo professionals perceive these activities as necessary and benign, they nonetheless constitute a form of power over animals, who are micro-chipped, filmed, transferred to other zoos and contracepted in order to regulate their reproduction. Power thus remains central to the modern zoo, even if its outward expressions and meanings have changed.

Knowledge

While exotic animal collections functioned, in part, as symbols of power, they also served as a venue for the production and dissemination of knowledge. Like museums, laboratories and botanical gardens, menageries provided a space for the study of the natural world, and also, in some instances, for its control and exploitation. How effectively they have fulfilled this role remains subject to debate.

By bringing live animals into human-controlled environments, zoos and menageries have offered a useful place for observing exotic beasts up close and studying their anatomy, means of locomotion and cognition. This has allowed naturalists to learn more about how animal bodies work and how animal minds function. Observing a captive giraffe in Calcutta Zoological Garden, for instance, Ram Brahma Sanyal noted that the animal had 'a rocking motion when walking', and was 'given to swinging its long neck every now and then, sometimes describing a complete figure of 8'.[29] When an infant chimpanzee named Tommy arrived at London Zoo in 1835, ZSL member William Broderip subjected the ape to a series of stimuli to test his reactions and instincts:

> When offered a glass of sherry, he raised it to his lips, but drank a very small quantity, as it did not seem agreeable to his taste. A cocoanut being presented to him, he threw it down several times successively upon the floor for the purpose of breaking it, anxious to get at its contents … [When] a python was brought in a hamper into the room … [he] manifested the utmost horror and aversion, retreating to his keeper, and evidently seeking his protection.[30]

The zoo environment was, of course, less suitable for evaluating interspecies interactions, natural foraging behaviours or hunting techniques, all of which were distorted by unnatural surroundings, atypical social groupings and abnormal diets. Nocturnal anteaters, for example, slumbered in their cages during the day, generating false conclusions about their 'slow movements' and 'stupidity of character'.[31] Despite these limitations, however, menageries provided basic information on animal behaviour and physiology, allowing naturalists and artists to gain an appreciation of the form and habits of previously unknown creatures. When zoo inmates died, moreover (a regrettably frequent occurrence in early collections), their cadavers provided welcome study material for comparative anatomists, who opened them up to examine their internal organs. In 1681 Joseph Guichard Duverney dissected an African elephant from Louis XIV's menagerie, learning more about the animal's skin, trunk and skeleton.[32]

Of more benefit to the animals themselves, menageries have contributed to knowledge about animal husbandry and veterinary medicine. In order to keep rare beasts alive, keepers have, of necessity, had to learn about their diseases, accommodation needs and dietary requirements. Early forays into these areas were founded largely on trial and error – with negative consequences for the animals concerned; an orangutan exhibited at the Jardin des Plantes in 1848 consumed 'chocolate, roast meat, wine and even liqueurs' and was 'put to bed between a large cat and a very shaggy dog' to keep him warm at night.[33] As time went on, however, understanding of animal nutrition and welfare advanced significantly, facilitating more sophisticated feeding plans, enclosure designs and veterinary treatment. In 1911, when a tapir at London Zoo developed a 'big abscess in the neck', the institution's resident veterinarian made a 'deep incision' in the growth to extract the puss and

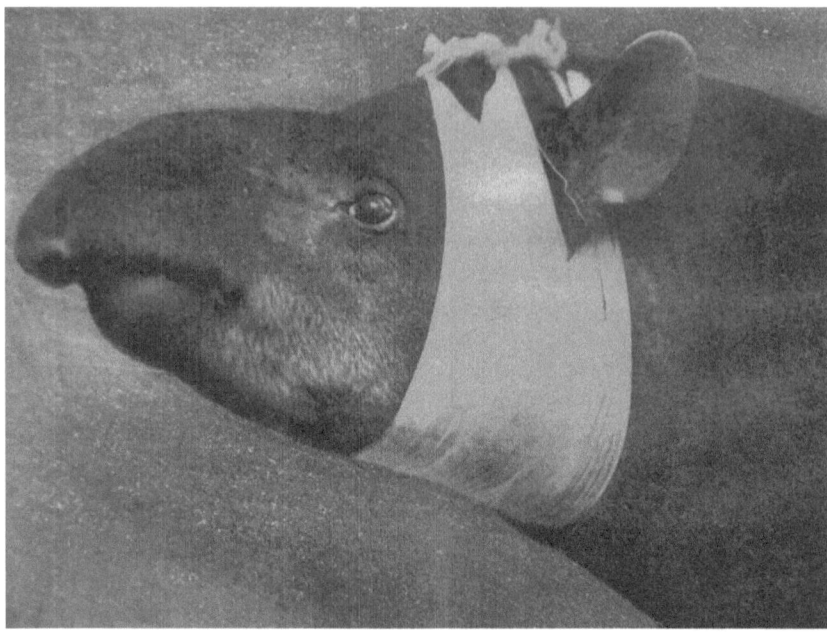

Figure 4.2 'American Tapir Suffering from a Swollen Face', from W.S. Berridge and W. Percival Westell, *The Book of the Zoo* (London: J.M. Dent and Sons, 1911), p. 23.

re-dressed the sore daily for three weeks to prevent infection (Figure 4.2).[34] In 1953, when one-year-old Indian elephant Melinda fractured her hind legs in a fall, director Robert Bean at Brookfield Zoo X-rayed the injured limbs to assess the damage, applied ice packs to reduce inflammation and placed the patient in a specially constructed canvas rig to facilitate healing.[35] Zoo animals have thus benefited from advances in surgery, antibiotics and anaesthesia, while also, on occasion, contributing to the development of human medicine. In the 1880s, for instance, surgeon John Bland-Sutton studied the causes of rickets in monkeys and lions at London Zoo to better understand the disease in human children.[36]

In the nineteenth century, zoological gardens contributed to another branch of contemporary science: the acclimatisation of exotic animals. Seen as a way of appropriating 'useful' species for consumption or aesthetic pleasure, acclimatisation constituted the central mission of several zoological institutions and was closely linked to wider economic ambitions. In France, an acclimatisation garden was established in the Bois de Boulogne in 1860, with the specific aim of breeding exotic animals.[37] In Britain acclimatisation was part of the original remit of the ZSL, whose founder, Sir Stamford Raffles, prioritised 'the introduction of new varieties, breeds and races of animals, for the purpose of domestication, or for stocking our farmyards, woods, pleasure gardens and wastes'.[38] Would-be acclimatisers hoped that non-native animals could be bred and raised locally, saving the state money on imported livestock and their products. Peruvian alpacas or Turkish angora goats, for instance, might be raised in France or Spain; Scottish salmon and English songbirds could offer sustenance and solace for British settlers in Australia. Despite some early successes, however, few acclimatisation projects produced tangible results, prompting most zoos to drop this goal from their portfolios by the turn of the twentieth century.[39]

More recently, the rise of environmentalism from the 1960s has seen the emphasis of most Western zoos shift from acclimatisation to conservation. Modern zoos have established breeding programmes to preserve endangered species. International stud books have been created to facilitate the exchange of animals for mating purposes and novel scientific techniques have been used to increase the chance of successful reproduction. In 1957 vets at Whipsnade Zoo injected Indian rhinoceros Mohan with 'a fantastically large quantity of gonadotrophic hormone' to encourage him to mate with his female counterpart Mohini – a tactic that resulted in the birth of the second rhinoceros calf to be born in captivity.[40] While zoos have had some conservation successes, however – most notably in preserving the California condor, Père David's deer and the Arabian oryx – serious questions remain over the ethics and efficacy of captive breeding initiatives. Is there any value in breeding animals which will probably never be reintroduced to the wild? Have zoos tended to prioritise the breeding of attractive and popular animals (such as pandas) over less cuddly species like reptiles and amphibians? Would artificially bred animals survive in the wild, even if they were released? And what should happen to surplus animals, whose genetics make them unsuitable for breeding? The genuine value of zoos as conservation centres is thus open to question, and their work in this area is seen by some as insufficient justification for their continued existence.

A fifth and final way in which zoos and menageries sought to contribute to knowledge was through public education. In early menageries, knowledge dissemination was limited and focused primarily on basic understandings of anatomy. Visitors might learn that the elephant had a trunk and the giraffe a long neck, but probably little else. By the nineteenth century, however, some menageries and most zoos had upped their game, employing signage, guidebooks and regular keeper talks to educate visitors and providing information about the scientific names, origins, habits and uses to man of the different animals on display. Zoo animals also featured in public lectures, and, from the twentieth century, on radio and television shows, bringing lions, tigers and chimpanzees to an audience beyond the zoo. In 1937 Julian Huxley exhibited an eight-month-old lion cub named Mac during an educational lecture to show that 'all lions, when young had spots, and that there was no adult spotted lion'.[41] From 1950 to 1957, Marlin Perkins, director of Lincoln Park Zoo in Chicago, hosted a weekly show called *Zoo Parade*, in which he introduced viewers across the USA to the institution's inmates. Such activities helped to familiarise non-expert audiences with the anatomy, physiology and behaviour of non-native animals, and more recently, to communicate the threats many now face in the wild.

What visitors learned from a trip to the zoo is less certain. Some undoubtedly absorbed information, acquiring knowledge they might not have gained through reading natural history books; a subscriber to Edinburgh Zoological Gardens cited as a praiseworthy example of such behaviour the case of 'a boy only four years of age, who, like most boys, had a picture book of natural history, and who, on being asked whether the representation of the dromedary was accurate, replied that it wanted the curtain over the eye, he having marked the long hair which protected the eyes of the specimen in the gardens'.[42] Other visitors, however, went to the zoo primarily to throw buns to the bears, ride on the back of the elephant or promenade in a fashionable setting, and probably learned little from their experience. Observing the behaviour of twentieth-century zoo visitors, Mullan and Marvin argue that 'unless there is some particular activity in a cage or enclosure, or unless the animal is a special favourite, it seems that, for the majority of people, watching consists of merely registering that they have seen something as they quickly move past it'.[43] The educational value of the zoo has thus been contested, and the degree to which intended messages about empire, animal behaviour or conservation make an impression on the average zoo visitor remains unclear.

Welfare

Zoos, menageries and circuses have also been studied from an animal welfare perspective. As sites where multiple wild animals are held captive, and, in some instances, forcibly trained to perform, menageries present several distinct concerns when it comes to animal cruelty. Institutional zoo histories tend to present a narrative of gradual improvement in welfare, culminating in the carefully designed, naturalistic enclosures of the modern zoo. While there are elements of truth to this argument (the twenty-first-century safari park is undoubtedly preferable to the nineteenth-century travelling menagerie), such a narrative is overly simplistic and conceals both the shortcomings of present zoos and the achievements of earlier generations. Moreover, as Nigel Rothfels has argued, it is questionable whether it is humans or animals who have derived most benefit from evolving modes of display. Have 'improvements' to enclosures been introduced to make the animals happy or, as Rothfels suggests, to make human viewers believe that they are happy?[44]

If we focus, first, on the exotic animal trade, we find a callous and brutal business that consistently put profit above welfare. Animal catchers routinely slaughtered adult females, or sometimes entire animal families, in their efforts to secure their more tractable cubs or calves. Captured animals endured long journeys overland and by sea and often perished from starvation. Changes of climate, meanwhile, and injuries sustained during capture, made wild animals susceptible to disease and often shortened their lives. Writing in 1872, British naturalist Frank Buckland estimated that, of 'six or seven rhinoceroses' caught in pitfalls in Malacca, 'the average is only one uninjured and fit to send to Europe'.[45] Sixty-five years later, *The Times* reported that two pandas captured in China had both succumbed en route to England, the male to a 'sceptic foot', likely 'damaged in a trap', and the female to the effects of being taken from her 'cold native heights to semi-tropical heat'. Since most of this violence happened out of public view, wild animal dealers were largely able to conceal the cruelties associated with capture and transportation and to downplay the collateral damage inflicted by their business. By the early twentieth century, however, public sensibilities were turning against the wild animal trade, and some raised their voices in opposition to it. Responding to a report of the two captured pandas, G.T.K. Maurice expressed horror at the prolonged suffering of the female, whom he pictured 'shaved of her thick coat and snugging herself down on a lump of ice to mitigate her misery and eventually dying of digestive disturbances after being starved to eat unnatural food. It is abominable that such things should be'.[46]

A particularly vivid example of the cruelties associated with the wild animal trade came to light in 1928, when a cargo of sixty Sumatran orangutans arrived at Cagnes-sur-Mer in the French riviera. According to a report in *The Times*, the animals had been snared using a new method of capture that involved 'concentrating a group into a clump of trees, which is then isolated by cutting down all of the surrounding vegetation', felling the trees in which the apes were sheltering and securing them 'by means of very strong nets'.[47] The orangutans had subsequently been sent by steamship to southern France, where they were awaiting dispatch to various European zoos – a process that would require a further gruelling overland journey. While the capture of great apes for exhibition in zoos was by no means unusual, the number of animals in the cargo was unprecedented, raising concerns as to both the suffering of the individuals and the impact of the trade on the survival of the species. The journalist who broke the story commiserated with the plight of one of the large males, who 'with an expression of profound grief, but quite gently ... submitted, for a bribe of a couple of bananas and a chunk of bread, to be lured into a contraption of sticks and wire, some 2ft. wide by 4ft. high by 6ft. or 7ft. long ... [t]hence ... into another and stouter enclosure, some 4½ft. square, in which ... he must stay till he gets to

Hamburg'.[48] Another observer, Sir Henry Hesketh Bell, equated the hunters of orangutans with 'the slave-raiders of old', and bemoaned the fate of the captive animals, who had once roamed 'in spacious liberty, through the huge forests, feeding on the fruits and succulent leaves of their choice', but were now 'cooped up in cages in which they cannot stand upright'.[49] The secretary of London Zoo, Sir Peter Chalmers Mitchell, also weighed into the debate, describing how two adult orangutans brought to London Zoo earlier the same year 'had the memory of their capture so vividly present in their minds that nothing would induce them to go through the open door of their house into the open-air exercising place which had been arranged for them, evidently thinking that the door was only the aperture of some kind of trap' – a reluctance that hastened their deaths from tuberculosis.[50] The coverage of the Cagnes orangutans thus vividly exposed the suffering inflicted by the wild animal trade and the inhumane methods used to catch and transport captive exotics.

While the wild animal trade was responsible for the most severe loss of life among exotic animals, survivors faced further forms of cruelty at the zoo or menagerie – some more visible than others. To begin with – and most conspicuous to the wider public – some menagerie inmates suffered physical abuse, either as a one-off occurrence or as a regular part of their training. In the ancient world and through to the early modern period, ruling elites often staged exotic animal fights as a form of entertainment. In the nineteenth century the more visceral violence of the organised fight receded, but it was replaced by the chronic violence associated with taming of menagerie and circus animals. Keepers frequently used coercive means to induce reluctant charges to execute unnatural manoeuvres, while animals that refused to perform were sometimes subjected to vicious punishments. Unruly or 'rogue' elephants were often poisoned or shot, especially males, which frequently grew dangerous and violent in captivity. Cruelty to performing animals behind-the-scenes was also a major concern for animal welfare advocates, who accused trainers of using excessive brutality in rehearsals. In 1908, for example, the RSPCA prosecuted elephant trainers William Schreida and his assistant Havaida for 'cruelly terrifying' an elephant during training by making it slide down a steep chute into a pool of water – a trick accomplished by digging 'the pointed end of [an] elephant stick six times into the right cheek of the animal' and prodding 'the hook into his back'.[51]

Overt physical violence was less common in nineteenth- and twentieth-century zoos, but more subtle forms of brutality often occurred out of public view. Animals that escaped, exhibited signs of violence or attacked a keeper, for instance, might be euthanised in the name of public safety, while individuals deemed to be diseased or surplus to breeding requirements might be culled – a practice that gained public attention in 2014 when Copenhagen Zoo announced that a healthy male giraffe named Marius was to be shot and fed to the institution's lions.[52] Though a less frequent occurrence, zoo animals have also suffered extreme physical violence during wartime, falling victim to human conflict. During the Paris Commune of 1871 many of the animals at the Jardin des Plantes were killed and eaten by starving Parisians, among them the zoo's beloved elephants, Castor and Pollux.[53] In World War II animals in several German zoos perished during Allied bombing raids, while the inmates of zoos across Europe (including London, Paris and Antwerp) expired from starvation or were slaughtered by their own keepers. Particularly distressing was the fate of the inmates of Ueno Zoo in Tokyo, who were systematically massacred in 1943 as the war turned against Japan. According to Ian Miller, most of the animals were poisoned with strychnine-laced meat, while the three elephants John, Tonky and Wanri starved to death in a grim ritual of total sacrifice.[54] Zoo animals have thus been subject to various forms of physical violence, whether as caged performers, surplus stock or collateral damage in war.

If physical (and public) abuse of captive animals has generated the most dramatic headlines, living conditions have presented a more chronic concern for zoos and menageries and continue to raise questions about the ethics of keeping wild animals in confinement. Often already traumatised by the process of capture and transportation, caged animals have suffered from poor diets, inadequate stimulation, unnatural social groupings and insufficient space, resulting in listlessness, lack of exercise, visible stereotypies and, in some cases, obesity and an early death. Others have succumbed to inappropriate contact from visitors, who have teased them, interrupted their sleep or supplemented their rations with unsuitable snacks. A polar bear deposited in the Calcutta Zoological Gardens in 1886 suffered profoundly from the intense heat of the Indian summer and survived for only a year and a half, despite being supplied with ice blocks to keep him cool.[55] A popular elephant seal named Roland perished at the Berlin Zoological Gardens in 1935 from 'injudicious feeding', his keeper having often found 'beer bottles, glasses and cigarette boxes in [his] tank'.[56] Generally worse in earlier periods, when understanding of animal physiology and behaviour was poorer, some of these bad conditions were addressed in the twentieth century through improved understandings of nutrition, more naturalistic enclosure designs and more suitable forms of enrichment, leading to better survival rates. In 1933 infant gorillas Mok and Moina at London Zoo were transferred to a newly constructed enclosure that included ropes and swings for entertainment, a water fountain that the apes could operate themselves and an 'air-conditioning plant which washes the air from fog, dust and germs'.[57] In 2002 Dr Mark Edwards at San Diego Zoo noticed that captive anteaters suffered from low energy and 'chronically loose stools' on their traditional zoo diet of 'meat, eggs, fruits, vegetables and biscuits', so substituted for this 'the meat-rich diet of a more commonplace carnivore, the domestic cat, with high-fibre monkey food added to replace the carbohydrate chitin, which forms the outer skeleton of ants'.[58] Such innovations have not been universal, however – nor always effective (both Mok and Moina died before they reached maturity) – and empty, sterile cages and inadequate animal husbandry have remained a feature of some zoos. Writing in 1976, Peter Batten described the dire conditions in dilapidated US zoos, where anteaters bruised their knuckles on concrete floors, big cats paced incessantly in small concrete cages and sea lions grew fungus on their bodies in dirty, non-saline water.[59]

How these issues should best be tackled remains a point of contention. Zoo reformers such as Batten have prescribed practical improvements and greater attention to the latest research in animal behavioural science as a solution to poor welfare in zoological gardens. Such a view has a long pedigree, and finds echoes in earlier protests about the treatment of zoo animals, which typically advocated larger cages and better animal husbandry. In 1906, for instance, a subscriber to *The Animal's Friend* praised Copenhagen Zoological Gardens for using 'toys' to improve 'the health and spirits of caged animals' – including 'a tub and a long rough pole' for the polar bears and a rocking horse for some of the monkeys.[60] More radical animal rights activists, on the other hand, have taken a harder line on the zoo as an institution, suggesting that captivity *per se* is a cause of suffering. Writing in 1841, following a visit to the Bristol Zoological Gardens, one commentator lamented in poetry the fate of the 'Caged Eagle … Bird of the far-commanding eye / and widespread wing' and asked 'who will not sigh, Thee coop'd and chain'd to see?'[61] In *When Elephants Weep*, meanwhile, Jeffrey Masson and Susan McCarthy contend that 'No cage is big enough for a polar bear or a cougar' and ask whether 'freedom to choose [is not] inextricably tied into the very notion of what it means to be happy?'[62] For these writers, freedom is a moral right, and not even the best zoos can ever fully replicate the native environments of their inmates. Such claims are supported by recent research in the field of animal behaviour science, which suggests that certain species fare badly in zoos and live shorter lives than their counterparts in the wild; zoo-born Asian elephants, for instance, live for an average of 18.9 years, while their counterparts born in Burmese timber camps (themselves not ideal conditions) live for an average of 41.7 years.[63] Where defenders of zoos tend to distinguish between good zoos and bad zoos, therefore, animal liberationists regard all zoos as cruel and call for their closure.

Global Animals

To what extent has the exhibition of exotic animals changed over time? Has the keeping of wild animals become more humane? Has the nature of the exotic animal trade changed? Have exotic animals elicited different responses in different eras and different cultural settings? To explore some of these issues in a little more detail, the chapter concludes with case studies of three global animals: a rhinoceros, a hippopotamus and a panda. Where did they emanate from? How did they end up in captivity? Who owned them and how were they received and perceived by the wider public? What can their stories tell us about continuities and changes in the exhibition of wild animals?

Ganda the Rhinoceros

Ganda the rhinoceros arrived in Lisbon in 1515. Captured in Surat in north-western India, Ganda (the Gujarati name for 'rhinoceros') was presented as a gift by the Sultan of Cambay to the Portuguese Governor of Goa Afonso de Albuquerque, who in turn gifted him to his sovereign Manuel I of Portugal. The rhinoceros travelled from Surat to Goa and from there to the port of Cochin in southern Indian, where he boarded the ship *Nossa Senhora da Ajuda*. He then journeyed by sea to Europe, rounding the Cape of Good Hope and reaching Lisbon on 20 May.

On his arrival in the Portuguese capital, Ganda immediately became a zoological wonder, attracting attention as both specimen and spectacle. Scholars studied the animal, noting his elephantine proportions, hog-like body and fabulous horn. Artists also sketched Ganda, recording his form for posterity. One of these sketches was sent to the celebrated engraver, Albrecht Dürer, who copied it to produce a famous woodcut of a rhinoceros. The image, which circulated widely, portrayed Ganda standing squarely on the spot, with curled lip, armour-like plates on his back, reptilian scales on his legs and an extra horn poking out between his shoulder blades.

As well as inspiring artistic representations, Ganda also became the protagonist in a violent animal combat. Keen to test the classical belief that the rhinoceros was the sworn enemy of the elephant, Manuel I organised a contest between Ganda and a young elephant from the royal menagerie. The combat took place in Lisbon's Terreiro do Paço on 3 June 1515. According to contemporary reports, the rhinoceros charged furiously at the elephant, which trumpeted in fear and bolted from the scene. Observers claimed that the rhinoceros attempted to gore the elephant's belly with his horn – as classical accounts had predicted.

After just seven months in Lisbon, Ganda was on the move again, this time to Rome, where he was destined to become a gift from Manuel I to Pope Leo X, and – possibly – to engage in a second combat with the latter's Asian elephant, Hanno. The rhinoceros never reached his destination, however. After a brief stopover in Marseilles, where he was paraded through the city dressed in a gilt harness and green velvet fabrics, Ganda perished in a shipwreck off the Ligurian coast. Some reports claimed that the rhinoceros's great bulk caused the ship to sink. Others stated that the pachyderm survived the wreck and attempted to swim to safety, but was dragged to his death by the chains that bound his feet.

Ganda's career encapsulates many key elements of early modern animal display. Originally from India, the rhinoceros arrived in Europe as a diplomatic gift from an Asian potentate, and was subsequently re-gifted to the Pope in another act of monarchical diplomacy. An 'exotic fetish', Ganda's exhibition at the court of Manuel I reflected a wider fascination in the Renaissance with oriental curiosities. A luxury commodity, his presence in Lisbon reflected Portugal's recent expansion into the Indian Ocean World, and its dominant role in the spice trade; in addition to the rhinoceros, the cargo aboard the *Nossa Senhora da Ajuda* included cinnamon, pepper, myrrh, sandalwood, aniseed, cloves and aloe – all coveted East Asian spices. Like many other exotic animals in this period, Ganda was not simply put on display in a cage but participated in combats and processions, sating a Renaissance passion for curiosities and gory spectacles. More benignly, the rhinoceros also became a cultural icon after his form was immortalised by Dürer, illustrating both the artistic advances of the period and, crucially, the power of the printing press, which allowed Dürer's image to be circulated widely (some 45,000 copies were sold within the artist's lifetime, shaping European ideas of what a rhinoceros looked like). Ganda's sad demise at sea highlighted the dangers of seaborne travel, foreshadowing the enormous casualties of the wild animal trade.[64]

Obaysch the Hippopotamus

Obaysch the hippopotamus was the first hippopotamus ever exhibited in Britain, and the first representative of his species to reach Europe since the Roman Empire. A native of East Africa, Obaysch was born some time in 1849 and captured by hunters near the island of Ghebbaysch on the White Nile, after which he was named. He was transported on camel back to Cairo, where he was cared for by the British consul, Charles Augustus Murray, and subsequently transferred to the port of Alexandria, where he boarded the P&O steamer *Ripon* to Southampton. During the voyage to England, Obaysch consumed 80 pints of milk daily and bathed regularly in a 'convenient iron tank, holding about 400 gallons of water'.[65] When he arrived at London Zoo he was enticed out of his transport van by his Arab keeper Hamet Saafi Cannana, who walked in front of him carrying a bag of dates. The ZSL attributed Obaysch's safe passage to Murray's 'watchful care, to the liberality of the Director of the Peninsular and Orient Steam Navigation Company, to the attention of Captain Moresby and the officers of the "Ripon" and to the faithful services of Hamet Saafi Cannana'.[66]

Following his arrival in Britain, Obaysch became an instant sensation. Thousands of visitors congregated at the docks in Southampton, hoping to catch a glimpse of the rare pachyderm. Thousands more flocked to see him at the zoological gardens, which experienced a spike in admissions. Scientists, artists and entrepreneurs also took a close interest in the hippopotamus, scrutinising his anatomy and physiology and using his striking image to promote their wares. The comparative anatomist Richard Owen subjected Obaysch to a close examination, describing his 'barrel-shaped' body, 'very short and thick legs' and 'thick hide'.[67] Don Juan Carlos María Isidro de Borbón, Count of Montizón, photographed Obaysch snoozing next to his pool (Figure 4.3), while the ZSL's chief illustrator, Joseph Wolf, painted the young hippo lying on his side, his short legs protruding from beneath his pink stomach. On a more overtly commercial level, Obaysch spawned a range of hippo-themed paraphernalia, including a small statuette (supposedly from Nile mud) crafted by the sculptor Joseph Gawen, a 'Hippopotamus Polka' composed by Louis Saint Mars, and a variety of hippopotamus-inspired trinkets. The novelist Charles Dickens saluted the hippo in his magazine,

Figure 4.3 Photograph of Obaysch. Juan de Borbón, *The Hippopotamus at the Zoological Gardens, Regent's Park* (1852). Salted paper print from glass negative. Image: 11.1 × 12 cm. Mount: 43.8 × 30.5 cm.

Source: Gilman Collection, Purchase, Ann Tenenbaum and Thomas H. Lee Gift, 2005. 2005. 100.14. www.metmuseum.org.

Household Words, praising the beast's 'unctuous appetite for dates, his jog-trot manner of going' and his 'majestic power of sleep'.[68]

Though challenged by other zoological 'stars' in the 1850s and 1860s, Obaysch remained popular throughout his life and was consistently one of the prime attractions at London Zoo; writing in 1877, the naturalist Frank Buckland claimed that 'no animal attracts more visitors at the Gardens than 'Hippo' when he is inclined to be lively' and 'We have seen persons fight even for the best places opposite his parade ground and bath'.[69] The old hippo made the headlines once more in 1872, when he sired a calf with the female hippopotamus Adhela (the animal, born on 5 November, was named Guy Fawkes, though it subsequently turned out to be female), and again in 1873 when he clashed with his mate in his den, 'grinn[ing] a ghastly grin, and loudly trumpet[ing] "Umph", "Upmh", "Umph"'.[70] When Obaysch died in 1878, he received an obituary in *The Times*, which imagined 'the fellows and friends of the Zoological Society' would 'hear with regret of the death of the old hippopotamus'.[71]

Obaysch's story exemplifies many elements of exotic animal exhibition in the Victorian era. Born in Africa, the hippopotamus's arrival in London was a product of imperial diplomatic networks and new transport technologies. A zoological novelty in Europe, Obaysch attracted the attention of the British scientific establishment and was studied intensively by contemporary naturalists, who observed his form and features and watched his behaviour in captivity. A public sensation, Obaysch also became a commercial product, to be marketed and consumed for financial gain.

Less loudly trumpeted in the contemporary press, but of equal importance, Obaysch's career illustrates the darker side of life of the exotic animal trade, which involved physical violence, psychological distress and frequent loss of life. Obaysch's mother was 'mortally wounded' by hunters during his capture, sharing the fate of many adult animals. He himself received a deep wound on his flank from a harpoon, leaving the scar visible on Montizón's photograph.[72] Once at the zoo, Obaysch likely suffered from nutritional deficiencies and cramped living conditions, and, according to several observers, grew increasingly bad-tempered and morose as he got older. Visiting in 1855, naturalist John George Wood witnessed the hippopotamus 'raise himself as high as he could from the water' after a visitor patted his nose, and 'with a furious snort' make 'such a plunge at the bars that the bystanders scattered off in fear'.[73] Obaysch's life thus encapsulates the trauma of capture and the frustrations of captivity as well as the glory of imperial connections and scientific advancement. Though keepers did their best to care for the hippo, enlarging his bath in 1855, installing 'a series of iron hot-water pipes' in his den to keep the 'temperature at a proper level', and even extracting a fractured tooth from his mouth with a pair of forceps, this care did not compensate for his loss of liberty, and could not replicate the conditions under which the animal would have lived in the wild.[74]

Tohui the Giant Panda

Giant panda Tohui was born in Chapultepec Zoo, Mexico City, on 21 July 1981. She was the second cub born to pandas Ying Ying and Pe Pe, two pandas presented to Mexico in 1975 by Chairman Mao Zedong. Measuring just 10cm long at birth, Tohui was kept in isolation with her mother for the first six months of her life and observed by keepers and the public on CCTV. This precaution was taken following the death of her older sibling, Xeng Li, who was suffocated by Ying Ying at just eight days old – an accident believed to have been caused by the stress of receiving human visitors.

As the first panda cub born outside of China to survive beyond infancy, Tohui was an immediate hit with the Mexican public, and, like Obaysch, boosted visitor numbers at the zoo. Some 100,000 people attended Chapultepec Zoo immediately following her birth (though police prevented their entry to the panda exhibit) and, once visits were permitted, 'a permanent queue of thousands and thousands of people' passed by her enclosure, trying to get a glimpse of the young panda. Panda memorabilia proliferated, including posters, dolls and a song by Mexican popstar Yuri. A nationwide competition to name the new arrival generated thousands of suggestions, with the winner announced on national TV. The winning entry, Tohui, means 'boy' in the language of the Tarahumara people (an indigenous group from northern Mexico) – an unfortunate choice, since the young panda, like Guy Fawkes the hippo, later turned out to be female (pandas are notoriously difficult to sex as their reproductive organs are internal).[75]

While Tohui entranced the Mexican public as a cub, her primary purpose as an adult was to breed, expanding the population of a highly threatened species. To this end, keepers at Chapultepec Zoo enlisted the latest knowledge and techniques available in panda procreation, securing her a mate, the male panda Chia Chia from London Zoo, in 1989, and carefully choreographing his arrival to ensure that she was in heat when he got there (a particular challenge in pandas, which are only fertile for 3 days per year). Advances in genetics and fertility science played an important part in the reproductive process, influencing the selection of a mate for Tohui, and, in 1990, allowing her younger sister, Xia-Hua to be artificially inseminated with sperm from another male panda. Keepers also paid close attention to the pandas' diet in captivity, feeding Tohui and her relatives on 'a liquid diet of eggs, milk, rice, apples, carrots, sugar and spinach with chicken or beef, all mixed in a blender', together with 'bamboo shoots for snacks'.[76] In July 1990 these efforts – assisted, according to some, by Mexico City's high altitude (comparable to that of the pandas' native Sichuan) – resulted in the birth of Tohui's cub, Xin Xin, the first offspring of a panda that had herself been born in captivity.

The experience of Tohui offers an interesting insight into the priorities of the modern zoo and the changes and continuities with earlier exotic animal exhibitions. On the one hand, in marked contrast to earlier exhibitions, conservation was the watchword in Mexico's panda breeding programme, and the Chapultepec Zoo justified its exhibition of the popular mammals on the grounds that it would publicise the plight of an endangered species. Tohui, a product of this breeding programme, was born in captivity, unlike Ganda and Obaysch, and contributed to a broader, transnational reproductive effort in which knowledge, experience and – in more recent years – panda semen, were shared between zoological institutions on different continents. She and her fellow captive pandas also benefited from the latest advances in nutritional, animal behaviour and veterinary science, whether in the form of food provision, behavioural enrichment, or medical treatment for illness; when Tohui's US counterpart, Ling Ling, suffered a kidney infection in 1983, staff at the zoo 'performed a biopsy and a kidney ultrasound, tested blood and bone marrow ... prescribed antibiotics' and borrowed a dialysis machine from a local hospital in case the drugs failed to effect a cure.[77]

In other ways, however, the panda project resembled older responses to exotic animals and perpetuated a longstanding fascination with charismatic mammals. Tohui's parents, Ying Ying and Pe Pe, arrived in Mexico City as diplomatic gifts, as part of China's increasingly sophisticated panda diplomacy. Tohui herself became a symbol of Mexican pride, as the first surviving panda born outside of China, while the outpouring of panda-themed commodities following her birth resembled the commercial response to London Zoo's first panda Ming, whose arrival in 1938 inspired 'panda dolls ... panda jokes ... children's books, nursery wallpaper, charms, bangles, cigarette cases [and] brooches'.[78] Despite access to better veterinary care than Ganda or Obaysch, moreover, Tohui died in 1993 from a bacterial infection aged just 12 – less than the 15- to 20-year life expectancy of pandas in the wild (though her daughter, Xin Xin, 34, is currently the longest-lived panda outside of China). While the facilities, practices and stated mission of the zoo had changed, therefore, some of its core characteristics have endured.

Conclusion

The careers of our three global animals highlight both changes and continuities in the exhibition of exotic animals. On the change side of the equation, exotic animals have become increasingly accessible to the public and conditions in captivity have generally improved over time. Princely menageries, open to a narrow elite, have given way to national zoos, accessible to a large swathe of the paying public. Modern zoos have moved away from the environmental destruction associated with the wild animal trade and embraced the rhetoric of conservation; Buenos Aires Zoo metamorphosed into a 'eco-park' in 2016, overseeing breeding programmes for tapirs and Andean condors'.[79] Animal husbandry has also improved considerably since the nineteenth century, with better nutrition, larger enclosures, improved hygiene and better veterinary care resulting in a longer life expectancy for zoo residents. Obaysch was around thirty when he died, which contemporaries considered a respectable age, but the oldest hippopotamus in captivity, a female named Bertha, died at Manila Zoo in 2017 at the grand old age of 65. Tohui enjoyed a significantly better diet than the first pandas at London Zoo, Ming, Sung and Tang, who subsisted on 'fresh corn stalks and apples'.[80] These changes align with the story of progress often related in institutional zoo histories and suggest that modern zoos have little in common with the menageries of the past.

While there have certainly been significant changes in the exhibition of exotic animals, however, the continuities across time are perhaps equally striking and offer some important caveats to the 'improvement' narrative. First, despite a shift in emphasis, zoos continue to exert power over animals, and some species (especially pandas) still function as diplomatic gifts. Second, while the range of animals on display has increased over time, audiences continue to gravitate towards a few charismatic mammals, with the result that many modern zoos still stock them for entertainment purposes. Elephants, great apes and big cats have almost always been popular, while reptiles, birds and less eye-catching mammals have generally drawn fewer crowds. Writing in 1911, for instance, *The Times* reported that '[a]mong the mammals [at London Zoo], excluding well-known beasts such as monkeys, giraffes and elephants, and the large carnivora, the kangaroos, swine and small rodents such as the coypu attract eager attention; deer are little noticed; antelopes are passed by with the formal remark that they are pretty; and strange creatures like the Tasmanian wolf, the wombat and the [red] panda awaken hardly any response in the public mind'.[81] Third, though staged animal performances may have fallen out of fashion in the twentieth century, the desire to see zoo animals being active, or to interact with them in some way, has not abated, as evidenced by the continued popularity of sea lion displays, feeding exhibitions and immersive enclosures, where visitors can enter the habitats of non-aggressive animals such as lemurs or marmosets. The chimpanzee's tea party of the early twentieth century (which persisted in London Zoo until 1972) was in many ways a sanitised version of feeding time at the Tower Menagerie in the sixteenth century, when visitors could reportedly gain free entry if they brought along a live dog or cat to feed to the lions! Finally, while living conditions for zoo animals have generally improved over time, this change has often come in response to changing human sensibilities (for example, an objection to seeing animals behind bars), rather than greater concern for animal needs, and zoo animals have continued to suffer from physical and psychological diseases that are absent in the wild. Carl Hagenbeck's much-praised barless zoo, for instance, patented at Stellingen in 1907, gave only the illusion of liberty for the animals, which were confined by moats and fake rocks instead of visible bars, and, in the case of birds, had their wings clipped to prevent them from flying away. Although the official ethos of the zoo has changed, therefore, the continuities between zoos and menageries across different historical eras are perhaps more striking than the differences.

Notes

1 Alan Mikhail, *The Animal in Ottoman Egypt* (Oxford: Oxford University Press, 2014), p. 111.
2 Eric Baratay and Elisabeth Hardouin-Fugier, *Zoo: A History of Zoological Gardens in the West* (London: Reaktion Books, 2002), pp. 17–19.
3 Bernal Díaz del Castillo, *The True History of the Conquest of New Spain*, trans David Carrasco (Albuquerque, University of New Mexico Press, 2008), pp. 130–131. For a detailed discussion of Moctezuma's menagerie and its wider significance see Matthew Restall, *When Montezuma Met Cortes: The True Story of the Meeting that Changed History* (New York: HarperCollins, 2018), pp. 117–150.
4 Randy Malamud, *Reading Zoos: Representations of Animals in Captivity* (New York: New York University Press, 1998), p. 64.
5 On exotic animals at court, see Annemarie Jordan Gschwend, *The Story of Suleyman: Celebrity Elephants and other Exotic in Renaissance Portugal* (Zurich: A Pachyderm Production, 2012) and Carlos Gómez-Centurión Jiménez, *Alhajas para Soberanos: Los animales reales en el siglo XVIII: de las leoneras a las mascotas de cámara* (Madrid: Junta de Castilla y León, 2011).
6 Edward Bennett, *The Tower Menagerie* (London: Robert Jennings, 1829), p.xiv.
7 'The Giraffe (Cameleopard)', *The Times*, 8 August 1827. See also Erik Ringmar, 'Audience for a Giraffe: European Expansionism and the Quest for the Exotic', *Journal of World History* 17:4 (2006), pp. 375–397.
8 Giuliano Bertuccioli, 'A Lion in Peking: Ludovico Buglio and the Embassy to China of Bento Pereira de Faria in 1678', *East and West* 26 (1976), pp. 223–240.
9 'Noticia del elefante remitido de Manila para el Rey nuestro Señor en la fragata nombrada Venus', in *Descripción del Elefante, de su Alimento, Costumbres, Enemigos e Instinto* (Madrid: Imprenta de Andrés Ramírez, 1773), p. 31.
10 Mikhail, *The Animal in Ottoman Egypt*, p. 127.
11 John Beusterien, *Transoceanic Animals as Spectacle in Early Modern Spain* (Amsterdam: University of Amsterdam Press, 2020), pp. 173–222.
12 Louise Robbins, *Elephant Slaves and Pampered Parrots: Exotic Animals in Eighteenth-Century Paris* (Baltimore: John Hopkins University Press, 2002), pp. 37–67; Peter Sahlins, *1668: The Year of the Animal in France* (New York: Zone Books, 2017), pp. 49–122.
13 'The Last Galapagan Tortoise', *The Standard*, 30 March 1898.
14 'Coming, Yes Coming!', *Atlanta Constitution*, 8 May 1890; 'Her Name is Clio', *Atlanta Constitution*, 10 August 1890. On zoos and civic identity see Andrew Flack, *The Wild Within: Histories of a Landmark British Zoo* (Charlottesville: University of Virginia Press, 2018); Elizabeth Hanson, *Animal Attractions:*

Nature on Display in American Zoos (Princeton: Princeton University Press, 2002); Lisa Uddin, *Zoo Renewal: White Flight and the Animal Ghetto* (Minneapolis: University of Minnesota Press, 2015).

15 Helen Cowie, *Exhibiting Animals in Nineteenth-Century Britain: Empathy, Education, Entertainment* (Basingstoke: Palgrave Macmillan, 2014); Peta Tait, *Wild and Dangerous Performances: Animals, Emotions, Circus* (Basingstoke: Palgrave Macmillan, 2012); Susan Nance, *Entertaining Elephants: Animal Agency and the Business of the American Circus* (Baltimore: Johns Hopkins University Press, 2013).

16 'A Remarkable Elephant', *The Atlanta Constitution*, 8 November 1898.

17 'A Day in the Life of a Trained Sea Lion', *Chicago Daily Tribune*, 15 September 1901.

18 'The Lions in a Fix', *Bristol Mercury*, 30 January 1858.

19 'Crimea', *Belfast Newsletter*, 27 October 1856.

20 Violette Pouillard, *Histoire des Zoos par les Animaux: Imperialisme, Contrôle, Conservation* (Ceyzérieu: Champ Vallon, 2019), p. 257.

21 'A Talk with Mr Carl Hagenbeck', *Manchester Weekly Times*, 15 December 1899; 'A Chat with Carl Hagenbeck', *The Era*, 11 May 1895.

22 B. Dashdulam, 'The Tale of Mongolia's Only Elephant', *The UB Post*, 24 July 2017.

23 Bob Mullan and Gary Marvin, *Zoo Culture* (Chicago: University of Illinois Press, 1987), pp. 49–50.

24 Marianna Szczygielska, 'Elephant Empire: Zoos and Colonial Encounters in Eastern Europe', *Cultural Studies* 34:5 (2020), pp. 789–810.

25 Pouillard, *Histoire des Zoos par les Animaux*, pp. 364–367.

26 E. Elena Songster, *Panda Nation: The Construction and Conservation of China's Modern Icon* (Oxford: Oxford University Press, 2018), pp. 84–101.

27 'Penguins to Fly North in Chartered Plane', *New York Times*, 13 March 1955.

28 Irus Braverman, *Zooland: The Institution of Captivity* (Stanford CA: Stanford University Press, 2013), pp. 187–188.

29 Ram Brahma Sanyal, *A Handbook of the Management of Animals in Captivity in Lower Bengal* (Calcutta: Bengal Secretariat Press, 1892), p. 153.

30 'The Chimpanzee', *The Times*, 29 October 1835, p. 3.

31 'The Zoological Gardens, Regent's Park', *North Wales Chronicle*, 29 October 1853.

32 Anita Guerrini, *The Courtiers' Anatomists: Animals and Humans in Louis XIV's Paris* (Chicago: Chicago University Press, 2015), pp. 202–204. On the pros and cons of studying science in zoos, see Miquel Carandell and Oliver Hochadel (eds), 'Science at the Zoo: Producing Knowledge about Exotic Animals', *Centaurus* 64 (2022), pp. 559–800.

33 'Zoological Curiosity', *The Times*, 14 September 1848.

34 'Operating on Wild Animals', *Sheffield Weekly Telegraph*, 23 November 1912.

35 'Two Broken Legs Put Baby Elephant in Sling', *Chicago Daily Tribune*, 4 October 1953.

36 Abigail Woods, 'Doctors in the Zoo: Connecting Human and Animal Health in British Zoological Gardens, c.1820–1890', in Abigail Woods, Michael Bresalier, Angela Cassidy, Rachel Mason Dentinger (eds), *Animals and the Shaping of Modern Medicine* (London: Palgrave, 2018), pp. 27–70.

37 Michael Osborne, *Nature, the Exotic and the Science of French Colonialism* (Bloomington: Indiana University Press, 1994), p. 11.

38 Thomas Allen, *A Guide to the Zoological Gardens and Museum* (London: Cowie and Strange, 1829), p. 5.

39 Takashi Ito, *London Zoo and the Victorians* (Woodbridge: Boydell and Brewer, 2014), pp. 138–161.
40 'First Indian Rhinoceros Calf Born at Whipsnade', *The Times*, 5 November 1957.
41 'Live Lion at a Lecture', *The Times*, 29 December 1937.
42 'Edinburgh Zoological Gardens', *Caledonian Mercury*, 22 April 1844, p. 3.
43 Mullan and Marvin, *Zoo Culture*, p. 133.
44 Nigel Rothfels, *Savages and Beasts: The Birth of the Modern Zoo* (Baltimore: Johns Hopkins University Press, 2002), pp. 201–202.
45 'A Cockney Rhinoceros', *The Times*, 10 December 1872.
46 'Captive Animals: Giant Pandas in England', *The Times*, 20 September 1937.
47 'Prisoners and Captives: Orang-utans from Sumatra', *The Times*, 15 May 1928.
48 'Apes at Wholesale', *The Times*, 24 April 1928.
49 'Prisoners and Captives: Orang-utans from Sumatra', *The Times*, 15 May 1928.
50 'Prisoners and Captives: Orangs at the Zoo', *The Times*, 21 May 1928.
51 'The Elephants at the Franco-British Exhibition', *The Animal World*, November 1908, pp. 245–247.
52 'Why did Copenhagen Zoo Decide to Kill Marius the Giraffe?', *The Guardian*, 9 February 2014.
53 'The Ethics of Dining', *London Society*, November 1883, pp. 550–563.
54 Ian Miller, *The Nature of the Beasts: Empire and Exhibition and the Tokyo Imperial Zoo* (Berkeley: University of California Press, 2013), pp. 120–162.
55 Sanyal, *A Handbook of the Management of Animals in Captivity*, p. 98.
56 'Death of Roland the Sea-Elephant', *The Times*, 31 December 1935.
57 'New Gorilla House', *The Times*, 31 December 1932.
58 'Cat Food for Aardvarks and other Zoo Diets', *New York Times*, 18 June 2002.
59 Peter Batten, *Living Trophies* (New York: Thomas Y. Crowell Company, 1976), pp. 39–72.
60 'Playthings for Caged Animals', *The Animals Friend* Vol XII, 1906, p. 35.
61 'To a Caged Eagle', *Bristol Mercury*, 4 September 1841.
62 Jeffrey Masson and Susan McCarthy, *When Elephants Weep: The Emotional Lives of Animals* (London: Vintage: 1996), p. 147.
63 Ros Clubb et al., 'Compromised Survivorship in Zoo Elephants', *Science* 322 (2008), p. 1649. Giraffes also have a shorter lifespan in zoos than in the wild, while cheetahs will not breed in captivity. For a discussion of how different species fare in a zoo environment, see Georgia Mason, 'Species Differences in Responses to Captivity: Stress, Welfare and the Comparative Method', *Trends in Ecology and Evolution* 25:12 (2010), pp. 713–721.
64 Juan Pimentel, *The Rhinoceros and the Megatherium: An Essay in Natural History* (Cambridge MA: Harvard University Press, 2017), pp. 15–118.
65 'Egypt', *The Times*, 20 May 1850. For a detailed account of Obaysch's career see John Simons, *Obaysch: A Hippopotamus in Victorian London* (Sydney: University of Sydney Press, 2019).
66 *Report of the Council and Auditors of the Zoological Gardens of London*, 1851 (London: Taylor and Francis, 1851), p. 14.
67 'The Hippopotamus at the Zoological Gardens', *The Times*, 6 June 1850.
68 'The Good Hippopotamus', *The Times*, 15 October 1850.
69 'Visits to the Zoological Gardens', *The Animal World*, May 1877, p. 69.
70 'Birth of a Hippopotamus', *The Times*, 7 November 1872; 'A Hippopotamus Fight in the Zoological Gardens', *Birmingham Daily Post*, 22 July 1873.
71 'The Zoological Gardens', *The Times*, 13 March 1878.

72 'The Hippopotamus at the Zoological Gardens', *The Times*, 6 June 1850.
73 J.G. Wood, *Sketches and Anecdotes of Animal Life* (London: G. Routledge, 1855), pp. 152–153.
74 *Report of the Council and Auditors of the Zoological Gardens of London*, 1855 (London: Taylor and Francis, 1851), p. 12; Wood, *Sketches and Anecdotes of Animal Life*, p. 166; 'Visits to the Zoological Gardens', *The Animal World*, May 1877, p. 70.
75 'La Familia de Pandas Mexicanas', *El Informador*, 10 August 1983; 'Ying-Ying Rompe Récord', *El Informador*, 26 June 1985.
76 'A Panda Needs a Lotta Love', *Chicago Tribune*, 11 October 1984.
77 'Will Love Triumph for the No.1 Celebrity at the National Zoo?', *Chicago Tribune*, 1 March 1984.
78 'Easter at the Zoo', *The Times*, 8 April 1939.
79 'From Showing Animals for Profit to Protecting Them: The Reinvention of Buenos Aires Zoo', *The Guardian*, 19 June 2024.
80 'Britain's First Giant Pandas', *The Times*, 23 December 1938.
81 'Popular Animals at the Zoological Gardens', *The Times*, 8 August 1911.

Further Reading

Baratay, Eric, and Elisabeth Hardouin-Fugier, *Zoo: A History of Zoological Gardens in the West* (London: Reaktion Books, 2002)

Braverman, Irus, *Zooland: The Institution of Captivity* (Stanford CA: Stanford University Press, 2013)

Carandell, Miquell and Oliver Hochadel (eds), 'Science at the Zoo: Producing Knowledge about Exotic Animals', *Centaurus* 64 (2022), pp. 559–800

Cowie, Helen, *Exhibiting Animals in Nineteenth-Century Britain: Empathy, Education, Entertainment* (Basingstoke: Palgrave Macmillan, 2014)

Flack, Andrew, *The Wild Within: Histories of a Landmark British Zoo* (Charlottesville: University of Virginia Press, 2018)

Hanson, Elizabeth, *Animal Attractions: Nature on Display in American Zoos* (Princeton: Princeton University Press, 2002)

Jordan Gschwend, Annemarie, *The Story of Suleyman: Celebrity Elephants and other Exotic in Renaissance Portugal* (Zurich: A Pachyderm Production, 2012)

Ito, Takashi, *London Zoo and the Victorians* (Woodbridge: Boydell and Brewer, 2014)

Malamud, Randy, *Reading Zoos: Representations of Animals in Captivity* (New York: New York University Press, 1998)

Mikhail, Alan, 'Enchantment' and 'Encagement', in Alan Mikhail (ed), *The Animal in Ottoman Egypt* (Oxford: Oxford University Press, 2014), pp.109–182

Miller, Ian, *The Nature of the Beasts: Empire and Exhibition and the Tokyo Imperial Zoo* (Berkeley: University of California Press, 2013)

Nance, Susan, *Entertaining Elephants: Animal Agency and the Business of the American Circus* (Baltimore: Johns Hopkins University Press, 2013)

Pimentel, Juan, *The Rhinoceros and the Megatherium: An Essay in Natural History* (Cambridge MA: Harvard University Press, 2017)

Pouillard, Violette, *Histoire des Zoos par les Animaux: Imperialisme, Contrôle, Conservation* (Ceyzérieu: Champ Vallon, 2019)

Ringmar, Erik, 'Audience for a Giraffe: European Expansionism and the Quest for the Exotic', *Journal of World History* 17:4 (2006), pp. 375–397

Robbins, Louise, *Elephant Slaves and Pampered Parrots: Exotic Animals in Eighteenth-Century Paris* (Baltimore: John Hopkins University Press, 2002)

Rothfels, Nigel, *Savages and Beasts: The Birth of the Modern Zoo* (Baltimore: Johns Hopkins University Press, 2002)

Sahlins, Peter, *1668: The Year of the Animal in France* (New York: Zone Books, 2017)

Simons, John, *Obaysch: A Hippopotamus in Victorian London* (Sydney: University of Sydney Press, 2019)

Szczygielska, Marianna, 'Elephant Empire: Zoos and Colonial Encounters in Eastern Europe', *Cultural Studies* 34:5 (2020), pp. 789–810.

5 Knowledge

How did the elephant get his trunk? Why do giraffes have such long necks? How do chameleons change colour? What is it like to be a bat?[1] These are just some of the questions humans have asked about other animals in different times and places, reflecting a longstanding curiosity about other species and a desire to explain how they live, how they think and how they have evolved over time.

As well as wanting to know about animals for their own sake, humans have also used animal bodies to garner information about themselves, employing other species as proxies for human beings. Guinea pigs, rats, mice, rabbits and dogs have served as test subjects for cosmetics and antibiotics. Armadillos have been infected with leprosy bacteria to develop a vaccine against the disease in humans. Baboons have served as living crash test dummies to determine the effect of high-speed motor vehicle collisions on the human body. These experiments have enhanced the length and quality of human lives – though at great cost to many animal lives.

This chapter explores the different kinds of knowledge that humans have produced in relation to animals and considers its wider cultural and ethical implications. The first part of the chapter examines the creation of knowledge about animals through the study of natural history. It charts the changing emphasis and scope of zoological research and assesses how new technologies – from the microscope to the camera trap – have shaped the ways in which we perceive other species. The second part of the chapter focuses on animal experimentation and traces its evolution over time. What experiments have humans performed on animals and why? What objections have been raised to the use of animals as experimental subjects? How suitable are animals as surrogates for humans? In bringing together these different strands of knowledge about animals, the chapter emphasises the key paradox at the heart of animal experimentation. In order to be valuable scientific subjects, the animals used in experiments need to be sufficiently close to humans to make the comparison valid. As our understandings of dogs, rats and chimpanzees reveal more about their sensory abilities,

DOI: 10.4324/9781003181996-6

cognition and capacity to experience pain, however, can we still justify using these animals as stand-ins for human beings?

Natural Histories

Humans have long been fascinated by other species and wanted to know more about them. What do animals eat? How do they move, hunt and breed? How do animals communicate? How do they comprehend the world around them? What, ultimately, differentiates animals from human beings? These questions – and others – have preoccupied humans for thousands of years, generating a plethora of texts, images, myths, fables and theories.

Interest in the forms and habits of animals dates back to the Palaeolithic era, when early humans first sketched the silhouettes of other species on cave walls. Around 1700–1200BC, humans living in caves at Altamira in northern Spain produced colourful paintings depicting bison, horses and wild boar. More than 12,500 years ago, humans at Serrania de la Lindosa in the Colombian Amazon sketched a range of native fauna onto rockfaces, including deer, armadillos, serpents, turtles, porcupines and a (now extinct) giant ground sloth. In the Drakensberg Mountains of eastern South Africa, the San painted lifelike images of eland (a large antelope), showing the animals being hunted with bows and arrows and honoured after death in dancing rituals (these paintings are difficult to date, but the oldest are believed to be around 2,400 years old). Though the purpose and significance of cave paintings is not known, their existence offers evidence of a longstanding curiosity about other species and a desire to observe them, represent them and record aspects of their behaviour.

The classical period witnessed more concerted efforts to observe animals, giving rise to the first recognisable natural histories – those of the Greek Aristotle and the Roman Pliny. Written around 350BC, Aristotle's *History of Animals* attempted to describe the key features of different species and, where possible, to explain how these came about. It divided animals into groups based on shared characteristics (birds, for instance, were placed in one group because they all had feathers) and provided detailed descriptions of their external and internal features. Viewed widely as a pioneering work in zoology, *History of Animals* drew in part on direct observation, in part on dissection and in part on the reports of travellers, fishermen and beekeepers.

Pliny the Elder's voluminous *Natural History* was written around 77AD and included four books on zoology. It drew heavily on Aristotle, but also added new facts and observations. Accorded access to Asian and African animals by Rome's imperial expansion, Pliny offered detailed descriptions of African beasts like the elephant and hippopotamus and mentioned more fantastical animals such as the phoenix. The Roman author also discussed the practical uses of animals, writing extensively about oyster farming, beekeeping and the murex snail, which served as a source of purple dye. Together, the works of Aristotle and Pliny set the agenda for many subsequent studies, shaping zoological research for centuries to come.

Observational natural history lapsed in medieval Europe, disappearing from the curricula of schools and universities. Animals remained central to medieval cultures, however, appearing on maps, in hunting manuals and in the margins of illustrated manuscripts. Marginalia in the fourteenth-century English manuscript known as the *Luttrell Psalter*, for example, depict a man herding a flock of geese, pigs foraging for acorns and dogs being set on a muzzled bear.[2] While traditional works of natural history were lacking in the medieval era, books known as bestiaries were widely produced in northern Europe, cataloguing all known species – from elephants to unicorns. Usually organised alphabetically, bestiaries provided the names and short descriptions of all known species, accompanied by illustrations. They focused primarily on the moral lessons that could be learned from different animals, using (sometimes fabulous) animal behaviours to illuminate aspects of the Bible. The pelican, for instance, was said to use the blood from its own breast to sustain its young – an example of maternal fidelity; the beaver symbolised chastity, because it bit off its own testicles to save itself from hunters.[3]

Beyond Europe, animals were likewise incorporated into the mythologies of a wide range of cultures, featuring in religious texts and creation stories. In Pre-Columbian Peru the Incas assimilated animals into their astronomy, identifying a llama-shaped constellation named *Yacana*, which was believed to drink the water from the ocean at night, and a constellation in the form of a snake, *Mach'acuay*, whose appearance marked the start of the rainy season.[4] In the medieval Islamic World several scholars pursued studies in zoology and veterinary medicine, drawing in part on classical authors. Ibn Qutayba (828–884) wrote a book of useful knowledge about animals, describing (accurately) how small birds would mob owls during the day, and claiming (with less plausibility) that horses had no spleen, camels no bile and ostriches no marrow in their bones. ʿAbd al-Laṭīf al-Baghdādī (1162–1231) collated information on Nile crocodiles. Al-Jahiz (776–860) is credited with writing the first known description of a food chain.[5] Like their European counterparts, these scholars drew upon a mixture of folklore and experience in their writings, blending personal observation with received wisdom.

The early modern period witnessed a renewed interest in the study of natural history, as naturalists sought simultaneously to recover the lost knowledge of classical authors and to study the previously unknown species that confronted them during the age of exploration. Buoyed by the desire to collect all the world's wonders into a single place, naturalists like Ulisse Aldrovandi in Bologna, Nicolas Monardes in Seville and Olaus Worm in Copenhagen gathered the physical remains of dead animals into curiosity cabinets or Wunderkammern, designed to inspire wonder in the beholder.

Faced with novel American species, like opossums, armadillos and sloths, travellers to the New World recorded information on their unusual forms and habits, attempting to convey their strangeness to audiences back in Europe. Describing the novel fauna of Peru, for instance, Spanish conquistador Pedro Cieza de León, characterised llamas as 'as large as small donkeys, with long legs, broad bellies … a neck of the length and shape of that of a camel' and 'large' heads 'like those of Spanish sheep'.[6]

Encounters with strange New World animals forced European naturalists to revise the way they studied natural history, moving away from their historic and symbolic associations and placing greater emphasis on empirical observation and visual representations.[7] This shift towards observation was complemented by a growing interest in dissection, which increased knowledge of the internal structures of other species. It was further facilitated by the invention of the microscope, which allowed scholars to observe the smallest anatomical features in vivid detail; in 1665, for instance, the British natural philosopher Robert Hooke looked at a flea under the microscope and sketched a magnified representation of the insect in his book *Micrographia*. While experiment and direct experience became increasingly central to understandings of animals, however, early modern natural histories were not entirely free from the more fabulous claims of the medieval bestiary, elements of which continued to feature in sixteenth- and seventeenth-century texts. The entry on 'The Bear' in Edward Topsell's *History of Four-Footed Beasts* (1607), for example, contained information on the animal's habits ('It has … been written that the old one frames the young ones to her own likeness with her tongue'); its moral qualities ('A bear is of a most venerous and lustful disposition'); and its uses to man ('If the blood or grease of a bear is set under a bed, it will draw unto it all the fleas and so kill them by cleaving thereunto').[8]

From the late seventeenth century, a new set of questions started to shape the study of natural history, with a particular focus on categorising and explaining difference. In the eighteenth century, the emphasis was on classification, as Enlightenment scholars sought to bring order to the natural world and better elucidate God's divine plan.[9] The Swede Carl Linnaeus for example, split animals into six different classes – mammals, birds, amphibians, fish, insects and worms – in his *Systema Naturae* (1735), basing his assessment on easily identifiable physical features (mammals, for instance, were distinguished by their trait of suckling live young). In the nineteenth century, naturalists started to ask more searching questions about the origin, dispersal and extinction of different animals, prompted by new developments in comparative anatomy and geology. How did new species come about? How could the presence of species in different continents be explained? Why and how did some species disappear? Some scholars, like the French comparative anatomist Georges Cuvier, posited that animal species disappeared from the earth as a result of dramatic geological changes (a process described as catastrophism). Others, led by another Frenchman, Jean-Baptiste Lamarck, proposed that traits acquired by an individual animal during its lifetime in response to environmental changes (for example a long neck, in the case of a giraffe) might be passed on to its offspring – thereby allowing species to survive and continually perfect themselves. A more sophisticated version of this transformist theory was put forward by the British naturalist Charles Darwin, who argued in his seminal *Origin of the Species* (1859) that natural selection – based on inherited (rather than acquired) traits could lead to gradual bodily changes that would, over several generations, enable species to adapt to environmental change (those that could not adapt would become extinct).

Underpinning debates over classification and evolution were wider historical developments that served to disrupt existing assumptions about the natural world. The 'discovery' of Australia in 1770, for instance, brought Europeans into contact for the first time with marsupials like the kangaroo and the koala. The exhumation of the fossilised bones of mastodons, giant ground sloths and enormous armadillos in North and South America brought the issue of extinction to the fore, compelling scholars to accept that the form and geographic range of different species might change over time, and that some might cease to exist. Classificatory debates were also influenced by longstanding questions about the human–animal divide, as naturalists struggled to articulate what (if anything) differentiated humans from other species. While some naturalists objected to the idea that humans might share a class with species as diverse as bats and whales (as suggested by Linnaeus), or a common ancestor with apes (as suggested by Darwin), others used spurious measurements of human body parts (skull volume, face shape, arm length) to argue that certain humans (non-Europeans, women, delinquents) were closer to animals than others (usually white men). Eighteenth- and nineteenth-century understandings of animals thus had significant implications for understandings of humans, and need to be viewed against the backdrop of exploration, racial segregation and colonialism.

The twentieth and twenty-first centuries have seen the increasing professionalisation of natural history, as amateur naturalists have given way to university-trained zoologists. They have also witnessed a simultaneous process of specialisation, marked by the emergence of newer, more niche disciplines such as entomology (the study of insects), ornithology (the study of birds), cetology (the study of whales and dolphins) and herpetology (the study of reptiles) – all focused on a specific category of animal. With the rise of ecology, and a growing emphasis on interspecies relationships, zoologists have started to undertake extended forms of field research, focused, increasingly, on understanding how animals shape their surroundings and interact with one another. In the early 1960s, for instance, British biologist Jane

Goodall spent three and a half years studying a group of 40 chimpanzees in the Gombe Stream Reserve on Lake Tanganyika, during which time she fraternised closely with individual animals and observed 'the rudiments of reasoned thinking, a complex social life and a fondness for blankets, of which they have stolen a great many from her tent'.[10] A parallel rise in ethology (the study of animal behaviour), has spawned a range of laboratory experiments designed to assess the cognitive capacities of other species, and to measure animal intelligence. In the 1970s and 1980s primatologist Francine Patterson attempted to teach American Sign Language to a female gorilla named Koko to see whether she was able to communicate with humans. In the early 2000s neurobiologist Keith Kendrick tested the memory and facial recognition powers of twenty sheep by showing them twenty-five pairs of sheep faces (one of which was associated with food), and repeating the experiment two years later to see if they remembered the earlier images (they did).[11]

Much of this research has focused on exploring the boundary between humans and animals, resulting, in some cases, in important reassessments of what (if anything) distinguishes humans from other species. Goodall's research revealed that 'chimpanzees will crumple up leaves to make a drinking sponge for water which they cannot reach with their lips' – a discovery that challenged the definition of man as the only tool-using animal.[12] Koko reportedly mastered around 500 signs and even invented her own gestures for things she had not previously been taught – 'red corn' for pomegranate, 'lettuce grass' for parsley and 'my cold cup' for ice cream.[13] These and other findings have demonstrated that what were once seen as some of the key defining features of being human – tool use, language use and the creation of culture – are not, in fact confined to humans, but shared by several other species.[14]

Places, People and Technologies

Where has knowledge about animals been produced? How, and by whom? How have new practices and technologies transformed the study of natural history? What ethical issues surround the collection and study of animals in the furtherance of zoological knowledge?

For many centuries, knowledge about animals was forged in two primary places – the museum and the field. Museums – often known in earlier eras as natural history cabinets – were spaces where remnants of dead animals were collated in the form of skins, bones and other body parts. They gathered specimens from regions, nations and empires, and enabled scholars to compare species from different parts of the globe. They also helped to communicate natural knowledge to a wider public (though access to these institutions varied considerably over time and between places).

The field encompassed a wide range of settings, from the savannah to the forest. Fieldwork (a modern term) entailed bringing the human observer to the habitat of the animal, rather than the other way around, freeing practitioners from the problems of poor taxidermy, inaccurate labelling and faded, moth-eaten skins. While museum-based scholars had the advantage of seeing their subjects alive, however, wild animals could be dangerous or elusive, limiting opportunities to examine their form, colours or habits. Questions have also been raised about the replicability and representativeness of observations in the field (was a behaviour observed among one group of lions typical of all lions, or unique to a particular group?), and, especially in the twentieth century, about whether animal behaviours witnessed in the wild might have been altered by the presence of the researcher, or by wider, human-induced changes to the natural environment.[15] For this reason, the testimony of travellers, sailors and even trained naturalists has often been dismissed as unfounded or fabulous – though the creation of permanent field stations from the early twentieth century improved the depth and credibility of in situ observations.[16]

In addition to museums and the field, anatomy theatres, menageries and, from the nineteenth century, laboratories, have provided additional places for learning about animals. Anatomy theatres and laboratories have given naturalists the opportunity to dissect dead (and sometimes living) animals to examine their internal organs. Zoos and menageries, meanwhile, have allowed zoologists to monitor individual animals over time and to assess their response to a range of stimuli, gaining important information about gestation periods or bodily changes in adolescence. The birth of several Malay tapirs in Breslau Zoological Gardens in the early 1900s, for instance, permitted the zoo's director, Fryderyk Grabowsky, to determine when and how the young animals lost their spots and stripes:

> [A]t the age of ten weeks the great white band, which the Germans call *Schabracke*, or saddle-cloth, is clearly visible, though the spots and stripes thereon may also be made out. Till the animal is between four and five months old the markings persist on the flanks and neck, but soon afterwards disappear, so that ... by the end of the fifth month the assumption of the adult dress is completed.[17]

Zoos, aquaria and laboratories have also served as places for conducting more systematic studies of animals, whether through subjecting animals to formal training or constructing experiments to measure their sensory or cognitive functions. This has enabled naturalists to observe behaviours that might not have been easily visible in the wild and to better understand how animals think, see, hear, touch and smell. In 1995, for example, a captive capuchin monkey astonished scientists at a research centre in Maryland by using a sharp flake of quartz to pry open a container of peanut butter, demonstrating the ability to make and use tools (a capacity previously believed to exist only in humans and other great apes, and later observed in wild capuchins in Brazil).[18] In 2006 scientists placed a huge mirror in the elephant enclosure at the Bronx Zoo to see whether the inhabitants were able to recognise their own reflections (a key marker of self-consciousness). They concluded that the pachyderms – three Asian elephants – did indeed associate the image in the glass with their own bodies, one of the animals, Happy, touching an X on her left cheek with her trunk after spotting the mark in her reflection.[19] A variety of settings have thus existed for acquiring knowledge about animals, all with different priorities, epistemologies and limitations.

If the place in which knowledge about animals was produced has influenced the nature and credibility of that knowledge so, too, have the people who produced it. Today, we generally view knowledge about animals as the preserve of professional zoologists. In earlier periods, however, the spectrum of people who acquired and shared knowledge about other species was much wider, ranging from farmers and hunters to soldiers, missionaries, sailors, merchants and colonial bureaucrats. Zookeepers, museum curators and artists all served as sources of information on animal bodies and behaviours, as, crucially, did many enslaved and indigenous people, who often acted as collectors of animal specimens and provided vital vernacular knowledge about their habits, medicinal uses and spiritual significance.[20] A painting by Jean-Baptiste Debret (1835), for instance, depicts a group of enslaved Brazilians collecting sloths, tropical birds and butterflies for European museums (Figure 5.1).

While a wide array of individuals contributed to the study of animals, however, not all knowledge was regarded equally and some actors have received more recognition than others. Jane Camerini shows, for instance, how the British naturalist Alfred Russel Wallace relied heavily on his Malay servant, Ali, while studying birds of paradise in New Guinea in the 1850s, but failed to reference the latter's contribution in any of his philosophical works (though he did mention him in letters and narrative writings).[21] Nancy Jacobs shows, likewise, how twentieth-century African bird experts like Jali Makawa, Saul Sithole and Njeru Kicho were excluded from certain scientific spaces and denied the pay and recognition accorded to their white collaborators.[22] Indigenous knowledge was thus not always granted the same status as European knowledge, and has often been appropriated without formal acknowledgement.

Gender has also been a significant factor in the production of knowledge about animals – and in the history of science more broadly. Until relatively recently, most zoological knowledge was produced by elite men, whose education and (often) wealth enabled them to engage in scientific pursuits. Women did, of course, possess knowledge about animals, and some contributed to

Figure 5.1 'Rétour à la ville des nègres chasseurs', from Jean-Baptiste Debret, *Voyage Pittoresque et Historique au Brasil* (Paris, 1834–1830).

Source: New York Public Library.

zoological texts, producing zoological illustrations or writing popular natural histories. With a few notable exceptions, however (perhaps most famously Dutch widow Maria Sibylla Merian, who travelled to Surinam in the early eighteenth century to study the country's insects), their work was not published under their own names and their input was unacknowledged.

Since the mid twentieth century the field of zoology has become more open to women, and female researchers have pioneered new areas of study. The primatologist Jane Goodall, for example, conducted pioneering work on chimpanzees in the 1960s, transforming the way in which great apes were studied in the wild. The zoologist Cynthia Moss carried out an extended study of elephants in Kenya's Amboseli National Park in the 1970s and 1980s, providing valuable insights into their lifecycles, social relationships and population dynamics. While women have become more prominent in the field of zoology, however, they remain under-represented in the discipline, and tend to be less cited than their male counterparts. Even in the twentieth century, moreover, gender stereotypes have sometimes influenced responses to female zoologists, generating reactions that would not have been directed at male scientists. One reviewer of Moss's book, *Elephant Memories*, criticised her use of human names such as 'Amy, Amelia [and] Alison' for her elephant subjects, which he felt 'smack[ed] of anthropomorphism'.[23] A 1964 article on Goodall described her, rather patronisingly, as an 'attractive and dedicated young woman' – adjectives that would almost certainly not have been applied to a male zoologist.[24] Knowledge about animals has thus often been gendered, as well as being influenced by the dynamics of class and race.

If place and gender have influenced the nature and credibility of zoological knowledge, change in that knowledge over time has frequently been driven by a third factor: technological innovation. This has significantly impacted what scholars are able to find out about animals and continues to shape new research in the field. If we think, first, about the study of animals in museums, advances in specimen preservation have had a major impact on what parts of animals could be exhibited and how long they survived. In the early modern period, only the more durable animal body parts could be preserved long-term, limiting the scope of study materials to fragments of animal bodies, such as teeth, claws, feathers, eggs and horns. Armadillos, for instance, were over-represented in early modern Wunderkammern because their tough shells survived longer than the furry bodies of other mammals.[25] From the early nineteenth century, however, arsenic-based compounds started to be used to preserve stuffed specimens, making them much more durable. The emergence of new modelling materials (clay, plaster, papier mâché) and new skinning and mounting techniques later in the Victorian era further improved the quality of specimens, triggering a shift from static rows of dead creatures to complex dioramas representing groups of animals in realistic postures against naturalistic backdrops. American zoologist Carl Akeley's tableau of a gorilla family at the American Museum of Natural History showed the apes disporting peacefully before a carefully recreated Congolese landscape – a much more naturalistic representation than would have been possible earlier in the century.[26]

Technological innovations were not confined to the museum, but also extended to the field. Beginning in the sixteenth century, and growing in prominence in the seventeenth and eighteenth centuries, zoological drawings and paintings assisted travelling naturalists in recording what they saw and making detailed representations of animals in their native landscapes (Figure 5.2). Technological advances (woodcuts, lithography and the advent of the steam press) in the nineteenth century made these images more accurate and easier to reproduce in large numbers, while the emergence of low-cost portable cameras in the 1890s enabled naturalists to photograph animals in the wild, and, later, to film them.[27] From the 1960s, furthermore, the invention of radio tracking technology permitted scientists to chart the movements of individual wild animals, gaining a more accurate understanding of their range and migration routes.[28] In 1970, for instance, scientists in South Africa implanted a tiny transmitter in the horn of a black rhinoceros called Rupert to track his food preferences, movements and activity patterns.[29] In recent years, these innovations have been supplemented by the use of camera traps, drones, satellite tracking and other innovative methods of monitoring

Figure 5.2 Anonymous, 'Otter, Sloth and Armadillo'. Dibujo. *Cuadrúpedos*. Bauzá Collection, Expedición Malaspina (1789–1794), drawing in pencil and ink.

Source: photograph by Gonzalo Cases.

wildlife in the field, allowing scientists to study animals at night or in dense jungle. In 2009 scientists studying Malay tapirs in Taman Negara National Park used camera traps to track tapir movements in the reserve, a procedure made possible by the discovery that 'the patterns of wrinkles on tapir's necks can identify individuals'. Their footage revealed that the density of animals in the park was much lower than initially assumed and showed that some tapirs regularly travelled 'up to three miles a night to reach salty mineral deposits'.[30] New technologies have therefore permitted naturalists to get closer to animals than ever before and to see them in new and revealing ways.

Lastly, it is important to address a question that has already been raised in the context of pet-keeping, animal exhibition and animal labour: what impact has the study of animals had on the creatures themselves, and what, if any, criticisms have been voiced about the exploitation of animals for the purpose of acquiring knowledge about them? Though a comparatively minor drain on wild animal populations, the study of natural history did, for many centuries, involve the killing of animals for research, something to which some contemporaries objected. Travelling naturalists and amateur collectors snared, darted and shot animals in the field and sent their remains to curiosity cabinets and museums for stuffing and display. Large numbers of beasts were often killed in this way, including rare ones. The American naturalist (and later prominent conservationist) William Temple Hornaday shot some of the last surviving bison in Montana so that he could exhibit them in the Smithsonian Museum. His compatriot, Charles H. Townsend, killed several of the last remaining northern elephant seals in California in the name of science.

Though not generally viewed as a major cause for concern in the Victorian era, hunting for the purposes of natural history started to generate critiques in the early twentieth century, as people began to worry about the environmental impact of killing rare species for museum display. These criticisms, while peripheral to the conservation movement as a whole, reflected a shift in emphasis, and a belief that the advancement of science was not enough to justify the killing of some animals. In one particularly emotive letter from 1934, for example, the Briton Albert Gray delivered a scathing indictment of zoologists who went to Africa to shoot gorillas for the proliferating number of museum collections in the US and Europe. '[T]he gorilla', he claimed, 'will never get due protection until the unholy alliance between museums and sportsmen is broken up'.[31]

Intentional killing of animals for zoological research has significantly decreased in the twentieth century, as new technologies have come on stream and public attitudes towards hunting have shifted. Some aspects of fieldwork, however, remain controversial, eliciting anger from animal rights activists. In the case of radio tracking, for instance, some animals have expired under anaesthetic while being fitted with radio collars, while others have died shortly afterwards, as a direct or indirect result of the procedure. A '370-pound female [grizzly bear] hit by a dart' during tagging operations in Yellowstone National Park in 1963 'staggered into a stream and collapsed' and would have drowned had scientists not interceded to drag her out.[32] Concerns have also been raised about the longer-term impact of attaching radio transmitters to animals, with critics suggesting that the weight of these devices may alter their behaviour (making them less useful to science) or make them more vulnerable to predators – or less able to catch prey – thus shortening their lives.

As for laboratory-based experiments, serious ethical questions remain about the treatment of animals used in behavioural studies, some of which have undergone painful or lethal procedures and many of which have been discarded after the research on them ended, spending the rest of their lives in zoos or being transferred to medical research programmes. Chimpanzee brothers Nim Chimpski and Ally, for instance, who gained fame in the 1970s for their mastery of American Sign Language, were subsequently sent to a laboratory in New York to test a vaccine for hepatitis.[33] Zoological research thus continues to pose ethical challenges for researchers, as scientists balance the desire to learn more about other species against the potential suffering of individual animals.

Natural Wonders and Living Fossils

How did evolving technologies, research questions and priorities affect the study of different species? To explore some of these issues in more depth, we will look at three species that long confounded human observers: the manatee, the platypus and the okapi. 'Discovered' by Europeans in different eras of imperial expansion, all three generated significant interest when news of their existence began to circulate. All three likewise generated inaccurate or misleading descriptions, giving rise to myths and misunderstandings.

The Manatee

The manatee (*Trichechus manatus*) is an amphibious mammal. Found in the West Indies, tropical South America, West Africa and Florida, it lives in both fresh and salt water, but surfaces to breathe at regular intervals. It has hand-like flippers, which it uses to propel itself along, and suckles live young. Though most closely related to the elephant and the hyrax, the manatee's resemblance to several other creatures has generated names reflecting its hybrid status; early Spanish colonists called the animal 'vaca marina' (marine cow), while in Portuguese it is known as 'peixe-boi', or bull fish. The English name, manatee, is derived from the Latin, 'manus' for 'hand'.

Europeans first encountered manatees in the sixteenth century. Baffled by the animal's hybrid qualities, they initially struggled to classify the species, making analogies with other, more familiar creatures. The Spanish Jesuit Bernabe Cobo, for instance, described the manatee as 'a fish native to the Indies', larger than a shark, with 'the head as large as an ox and somewhat similar in appearance' and flesh that tastes like 'veal or fresh pork'.[34] British explorer Robert Harcourt, who saw West Indian manatees in Guiana, observed that 'we account [the manatee] halfe flesh, for the bloude of it is warm', and, though a 'sea-fish', 'in taste, it is like beefe, will take salt and will serve to victual ships'.[35] Considered by turns a cow, a fish and even, by some, the physical embodiment of a mermaid (a belief fostered by the manatee's habit of sitting vertically in the water), the manatee reflected the challenges of fitting New World animals into classical and biblical frameworks and highlighted the unreliability of travellers, who often embellished what they saw. Reception of the animal also reflected the ways in which practical priorities could, on occasion, determine classification choices; valued for their skin, meat and fat (used for cooking), manatees were classed as fish by Spanish colonists, allowing humans to consume their flesh on Catholic fast days.[36]

In the eighteenth and nineteenth centuries scholars made further efforts to study the manatee, assisted, first, by the arrival of stuffed specimens in Europe, and later, by the study of living animals in captivity. In 1867, for instance, British zoologist James Murie dissected a dead manatee to learn more about its internal organs.[37] In the early 1880s Agnes Crane studied two captive manatees at the Brighton Aquarium, noting their unusual feeding habits:

> Notwithstanding the predilection they have evinced for land vegeta-bles, they never feed out of water. Food has been repeatedly offered them, but it always remains untouched, although readily devoured when the influx of water sets the leaves floating on the surface.[38]

Closer analysis of both living and dead manatees allowed naturalists to glean new information about their anatomy and physiology and to classify them more accurately (Figure 5.3). Detailed knowledge of the manatee's form and behaviour nonetheless remained patchy, hampered, on the one

Figure 5.3 W. Elliot, Plate 33, 'The Feeding Manatee', from George Brown Good, *The Fisheries and Fishery Industries of the United States* (Washington, DC: Government Printing Office, 1884).

hand, by the poor condition of museum specimens and, on the other, by the sickly constitution of most captive animals. The skin of a manatee sent to the Real Gabinete in Madrid in 1779, for example, arrived 'quite badly maltreated and lacking the head of the amphibian', forcing director Pedro Franco Dávila to improvise a head based on written descriptions.[39] The fifth manatee exhibited at New York Aquarium in 1906 sustained 'a spear wound at the base of the tail' during capture, which required 'careful dressing' each day with antiseptic. Its four predecessors all succumbed to pneumonia within months of their arrival – a fate keepers hoped to avoid for 'Manatee V' by heating the water in his pool to 27°C and covering it with a canvas screen every night.[40]

The twentieth and twenty-first centuries have brought new tools to the analysis of manatees, giving scientists novel insights into their range, behaviour and – crucial from a conservation perspective – the threats they face from human beings. In the field, aerial surveys have allowed scientists to estimate manatee populations in southern Florida, using the scars on the animals' bodies (sustained through collisions with boats) to identify individuals. Scientists have also employed radio tracking to monitor manatee migration patterns, with some surprising results; in the early 1990s a male manatee named Chessie astonished scientists by swimming north up the Atlantic coast on three consecutive years, reaching Chesapeake Bay in 1994 and Rhode Island in 1995.[41] In the laboratory, meanwhile, tests on captive manatees have given scientists a better understanding of the species' sensory capacities, undermining longstanding perceptions of manatees as slow and stupid. In 2006, for instance, neuroscientist Roger Reep analysed manatees Hugh and Buffett at the Mote Marine Laboratory, Sarasota, and established that their eyes could detect colour and form, that their hearing was sharpest at higher frequencies (10–20kHz) and that the vibrissae (whiskers) around their mouths gave them an excellent sense of touch, equal to the human hand or the elephant's trunk.[42] These findings challenged older assumptions about manatee cognition and intelligence.

Changing human responses to the manatee thus reveal both the shifting focus of natural knowledge and the degree to which wider cultural, social and economic priorities have impacted on the classification and presentation of other species. A 'warm-blooded, air-breathing, milk-giving, plant-eating, harmless water animal' – to quote one early twentieth-century American journalist – the manatee straddled the boundary between mammal and fish, eliciting different responses from different human actors.[43] For impressionable sailors, manatees were mythical creatures and the inspiration of fantastical reports of mermaids. For indigenous peoples and early European settlers, they were a source of food, which influenced their classification as 'fish'. For eighteenth- and nineteenth-century naturalists they were a taxonomic challenge, to be solved through empirical observation of living and dead specimens. And for twentieth- and twenty-first century marine biologists they were (and are) an endangered species, to be monitored, studied and dissected to gain knowledge with which to ensure their better protection from human threats (see Chapter 6). These shifting priorities have shaped the questions that have been asked about manatees – and the answers they have generated.

The Platypus

While the manatee presented a challenge to naturalists, the platypus (*Ornithorhynchus anatinus*), was a true zoological enigma. Furry and warm-blooded, platypuses suckle their young with milk secreted in the glands of the abdomen. They also possess duck-like beaks, which they use to detect shrimps and worms, and lay eggs in burrows, like reptiles. Male platypuses have a venomous spur on the heel that releases a strong poison, capable of causing paralysis and excruciating pain.

Long known to Australia's aboriginal people, platypuses were first encountered by Europeans in the late eighteenth century, following Captain Cook's 'discovery' of the southern continent. Like manatees, they baffled early settlers with their strange combination of mammalian and avian features, generating a series of jigsaw-puzzle descriptions. The naturalist George Shaw, who published the first scientific description of the platypus in 1799 (based on a stuffed specimen), described the animal as having a beak like that of a duck, a body like that of an otter and fur like that of a beaver.[44] Three years later, the surgeon Everard Home dissected the bodies of a male and female platypus, providing a detailed account of the animals' external and internal features – notably the absence of teats in the female, the 'strong, crooked spur' on the foot of the male and the platypus's distinctive bill, which he considered to be 'capable of nice discrimination in its feeling'.[45] Both naturalists struggled to place the platypus within existing systems of classification, debating whether it was a mammal or a bird. Shaw, indeed, initially doubted the authenticity of the specimens he received, suspecting they might have been assembled from the body parts of different animals.

By the middle of the nineteenth century the platypus's existence had been broadly accepted, but the species remained a source of scientific interest on account of the light its existence might shed on wider questions of evolution. In 1832 the comparative anatomist Richard Owen dissected five dead female platypuses to determine whether or not they possessed mammary glands (a key mammalian characteristic), noting that, when the suspected glands of one specimen were compressed, 'there escaped from these orifices minute drops of a yellowish oil, which afforded neither perceptible taste nor smell'.[46] In 1884 naturalist William Hay Caldwell made the 'discovery' that female platypuses laid eggs rather than birthing live young, reinforcing the idea that the platypus was on the boundary between birds and mammals. More detailed information about platypuses, both in the museum and in the wild, allowed naturalists to develop a better understanding of the species, fuelling speculation over the creature's ancestry – was it a kind of missing link between birds and mammals? Western knowledge of the species remained, nonetheless, limited in this period, held back by the challenge of observing platypuses in their native environment, the impossibility of transporting live platypuses to overseas zoos and the poor quality of Victorian taxidermy. Writing in 1897, naturalist Frederick Aflalo complained that 'little of the true character of this beautiful, glossy creature is to be learned from the miserable effigies, rigid as mummies, with the shrunken feet and brittle bill, presented for instruction in museums'.[47]

As with manatees, the twentieth century witnessed an increased focus on the study of living platypuses, first in the zoo, later in the laboratory. In 1944, Jack and Jill, two platypuses at Healesville sanctuary for Australian fauna, bred successfully in captivity, allowing curator David Fleay to observe Jill's nesting behaviour and to estimate the incubation period for her eggs (he guessed six days).[48] In 1947 platypuses Cecil, Penelope and Betty survived a trip by plane, boat and train to the Bronx Zoo in New York, enabling scientists to closely monitor their diet and behaviour. Betty perished after a short time, but Cecil and Penelope lived for ten years in captivity, revealing what zoologists believed to be a fast metabolism (the animals 'consume[d] their own weight in worms, crayfish and the occasional frog' in just 24 hours), a 'high tension temperament' (keeper Mr Crandall reported that 'no-one … has been able to hold … Penelope or … Cecil for more than a few seconds') and an apparent 'low' intelligence (a conclusion derived from the fact that 'they had never shown recognition of employees who had been feeding them').[49] Following the premature deaths of three subsequent captive platypuses in 1959, Australia banned further exports of the live animals, putting an end to their study outside of their homeland. The analysis of platypuses continued in Australia itself, however, drawing on a range of new technologies. In 1998, scientists at the national University in Canberra explored the possibility of developing a new generation of painkillers using the poisons produced by male duck-billed platypuses.[50] In 2008 an international team of geneticists analysed the DNA of a female platypus named Glennie from Glenrock Station, New South Wales, and compared it to that of several other species. They discovered that Glennie's genome contained around 18,500 genes, some with links to mammals (antibacterial proteins and genes associated with lactation) and some more commonly found in reptiles (those for egg-laying, vision and venom production), underlining the platypus's hybrid status.[51]

The case of the platypus illuminates the ways in which the study of natural history has evolved over time. In the eighteenth century, the emphasis was on describing the anatomy of a novel animal and fitting it within the latest classificatory schema. In the nineteenth century, the focus shifted to understanding how platypuses had come into being and how they related to other species. In the twentieth century, the presence of the first platypuses in zoos offered opportunities to study platypus behaviour and breeding habits. In the twenty-first century, zoologists have begun to re-visit questions of classification and evolution from a new angle, drawing on novel techniques such as DNA analysis.

As with the manatee, the platypus story also shows the confusion elicited by this unusual species, the challenges of studying and preserving the animal – either alive or dead, and the mistrust of indigenous knowledge, which was often far in advance of Western science (Aboriginal people were the first to report that the platypus laid eggs, but their testimony was dismissed as 'ignorance' by some European naturalists).[52] More disturbingly, the history of the platypus reveals a longstanding tendency to view the species in a negative light, either as a strange monster, composed of the body parts of other animals, or, later, as a bizarre relic of the past, preserved only because of its isolation from potential predators or competitors. As late as 1947, for instance, one journalist for the *New York Times* characterised the captive platypus, Cecil, as 'a bit on the doltish side', in reference to his apparent stupidity.[53] This negative perception of platypuses – and marsupials more widely – has been characterised by the historian Robert Paddle as 'placental chauvinism' and reflects a Eurocentric tendency to present what is different as inferior.[54]

The Okapi

The okapi (*Okapia jonstoni*) is a large, herbivorous mammal and the closest living relative of the giraffe. A native of the Ituri forest in the Congo, the okapi is notable for its thick oily fur, which helps it to stay dry in rain, and for the scent glands on the bottom of its hooves, which enable it to mark its territory. Like its better-known relative, the animal has short horns covered in skin, a prehensile tongue for grasping foliage and an elongated neck. It is predominantly black in colour, with white stripes across the rump to camouflage it from predators.

Unknown to Europeans until the early twentieth century, the okapi was first 'discovered' by British explorer Sir Harry Johnston, who, in 1900, secured 'a complete skin and two skulls of this animal' for the British Museum.[55] Johnston, who never saw a living okapi, believed that the creature was a new species of zebra. Naturalist Edwin Ray Lankester at the British Museum, however, concluded that it was 'a giraffe-like creature devoid of horns, with relatively short neck and with colour stripes on the limbs, but nowhere showing spots or areole like those of the giraffe'.[56] Often viewed, like the manatee and the platypus, as a kind of 'jigsaw-puzzle' animal, the okapi was described as having the 'shape and general body-colouring' of a hartebeest (a species of antelope), the head of a giraffe and the 'curious horizontal stripings upon the forelegs and hind-quarters' of a zebra.[57] Also like its fellow animal novelties, the species received negative coverage from some commentators, who perceived it as strange, ugly and atavistic – a kind of 'living fossil'. Writing in 1901, one journalist referred disparagingly to the okapi's 'drooping, ungainly' tail, 'preternaturally quick hearing', 'bizarre stripiness', and 'protruding and unlovely' eye, suggesting that the species 'may perhaps, like the tapir, be looked upon as a singular and bizarre instance of arrested development'.[58]

Following Johnston's initial 'discovery', information about the okapi emerged only gradually, owing to the difficulty of accessing the species. In 1905, M. Jeannet, 'a young Swiss official of the Congo Government', became the first European to see an okapi in the flesh after he was alerted to 'a beast standing at the side of the forest track' by a 'native soldier' (he immediately shot the animal for a museum in Brussels).[59] In the early 1930s, meanwhile, the Italian explorer Attilio Gatti conducted the first extended study of the species, photographing okapis in the wild for the first time and capturing a young male okapi named Toto, whom he fed on goat milk and lettuce leaves (Figure 5.4).[60] While understandings of the okapi slowly deepened, however, the okapi's inaccessible habitat and reclusive nature limited scientific study, and it was not until the late twentieth century, with the aid of new tracking technologies, that some aspects of its behaviour came to light. A twelve-year study (from 1985 to 1997) by John and Therese Hart, for instance, used radio collars to chart how much territory individual okapis occupied and what plants they ate. It revealed that okapis are 'intensely private, getting together only at mating times', that mothers hide their calves 'away in deep underbrush' so that they can 'wander part of the day foraging for food', and that calves don't defecate at all for the first two months of their lives 'lest the odour attract a lurking leopard'.[61]

As with the manatee and the platypus, the difficulty of catching and transporting live okapis for exhibition in zoos slowed the process of studying them – while adding to their allure. The first living okapi to reach Europe was presented to Antwerp Zoo in 1919 by Mme. Landaghem, who hand-reared the animal 'at her husband's quarters, several miles above Boma, on the Congo'.[62] The first okapi in the USA arrived at the Bronx Zoo in 1937, surviving the long transatlantic sea-crossing on a 'daily menu' of 'six litres of milk, nine bananas and three heads of cabbage'.[63] Usually snared in pitfalls, okapis were notoriously hard to catch and often sustained injuries during capture that brought about their death; of the three okapis caught alive by Gatti, one died immediately from shock, a second escaped his captors, while a third, Toto, succumbed to intestinal disease after just two weeks in captivity.[64] This situation only changed in the 1950s, when Portuguese

Figure 5.4 'He projected a good foot of tongue.' Toto, from Attilio Gatti, *Great Mother Forest* (London: Hodder and Stoughton, 1936).

Angolan Jean de Medina established a special okapi breeding station in the Congo from which the animals were exported around the world for high prices (a male sold for $5,000 and a female for $6,000).[65] Premature deaths of captive okapis were also common, though survival rates improved as zoo keepers learned to simulate the okapi's native diet and surroundings. In 1937 keepers at London Zoo used 'a carpet of peat', 'shady trees' and 'bamboo shoots' to make the outdoor enclosure of the institution's second okapi 'more nearly resemble the forest floor to which the animal is accustomed'.[66] In 1959 keepers at Copenhagen Zoo inserted 'green glass' in the okapi house to 'reduce the light to the dimness of the jungle'.[67]

The okapi story highlights the difficulty of studying a solitary, reclusive species and the crucial role of indigenous people as sources of zoological knowledge. Rarely seen by Europeans, okapis would likely have remained unknown to them without the testimony of indigenous Africans. African expertise, however, brought okapis to Europeans' attention and was vital in securing the first zoological specimens living specimens. Johnston, for instance, quizzed 'a party of dwarfs [sic] that inhabit the Semliki forest' about the okapi and obtained from them 'some pieces of its skin'; another British explorer, Percy Powell-Cotton, obtained details about the animal's acute hearing and grooming habits from the Mambutti.[68] Indigenous people were thus central to European understandings of the okapi, helping Europeans to view, understand and capture it. Their contribution to science was, nonetheless, often minimised, and their relationship with the animal branded as superstitious or environmentally detrimental; the Mambutti viewed okapis primarily as a source of okapi belts (important status symbols for warriors), and were accused of pushing the animal to extinction through over-hunting.

The okapi saga also illustrates the importance of place in shaping knowledge about animals. In the early twentieth century, the scant knowledge available about the species caused tensions between scientists working in the museum and scientists operating in the field, as the former questioned the credibility of the latter – and vice versa. Debating the range and size of the okapi population, for instance, Gatti tussled with Guy Dollman of the Natural History Museum, adducing his extensive fieldwork to corroborate his own (contested) claim to have discovered a new subspecies of okapi:

> Captain Dollman may rest assured that before announcing this new race of okapi, I not only made detailed inquiries, but examined carefully 14 live okapi, 137 skulls belonging to specimens of both sexes and races and of every age, and even three complete foetus – material, I dare say, that not all the museums and zoos of the world have at their disposal, and that only two years in the interior of the forest and the friendly confidence of natives and pygmies [sic] allowed me to see.[69]

More recently, by contrast, okapi research programmes have involved close cooperation between zoos, museums and field naturalists, which often collaborate to research and conserve the okapi. The Harts' intensive study of okapis in the Congo was funded by Chicago's Brookfield Zoo, which, in turn, adjusted its treatment of captive okapis in light of their findings, separating its five animals into different habitats to reflect their naturally solitary behaviour. The changing ways in which the okapi has been studied thus illustrates the increasing role of knowledge exchange in zoological research, as well as the increasing representation of women in science.

Experimenting on Animals

So far, we have focused on the acquisition of knowledge about animals for its own sake, to further human understandings of animal forms, behaviour and evolution. As well as wanting to know about animals from a natural history perspective, however, humans have also enlisted animal bodies to serve more directly anthropocentric ends. This has meant conducting experiments on non-human animals to gain a better understanding of human anatomy and physiology.

Animal testing has played a vital role in the advancement of human medicine, helping humans to combat infectious diseases, survive exposure to poisons and develop an increasingly sophisticated array of vaccines, antibiotics and surgical techniques. It has also, however, generated considerable opposition on the grounds of both welfare and efficacy. The second part of this chapter outlines some of the key areas to which vivisection has contributed and discusses the ethical challenges posed by animal experimentation. Is animal testing cruel? Has it been effective? To what extent have the benefits justified the costs?

Animal experimentation has encompassed a wide range of areas, providing humans with information about the form, function and degeneration of the human body, and its likely response to a range of stimuli. These can be subdivided into four distinct (though related) categories: (1) understanding the workings of the healthy human body; (2) understanding the cause and course of human ailments; (3) preventing disease through the development of vaccines; and (4) treating diseases with medicines or, in some cases organ transplants.

The earliest examples of animal experimentation focused on dissecting and vivisecting animal bodies to learn more about human anatomy and physiology. In ancient Rome, the Greek physician, Claudius Galen (129–216AD) vivisected pigs, goats and Barbary apes to illuminate the workings of the nervous system. In seventeenth-century Britain the physician William Harvey tied ligatures around the veins and arteries of rabbits to prove that blood circulates throughout the body, rather than being absorbed by the organs and constantly re-made in the liver. Half a century later, Harvey's compatriots Robert Boyle and Robert Hooke placed mice, cats, birds and snakes in a vacuum pump to prove that animals required fresh air (as opposed to old, respired air) in order to breathe. Often prevented by religious dictates from dissecting human cadavers, early anatomists relied on studying the animal body to illuminate the workings of its human equivalent. Dissection and vivisection thus played a key role in revealing the form and function of different human organs, giving scholars an insight into the circulatory, respiratory and nervous systems.

Animal experimentation has also been used to assess the impact of toxins on the human body and, more recently, to trace the progress of common human diseases such as cancer. In the early nineteenth century, as Victorians fretted about the physiological effects of various poisons, toxicologists dosed animals with venomous substances to assess their toxicity. The Spanish chemist Mateu Orfila, for example, performed extensive (and gruesome) experiments on dogs to determine the impact of arsenic on the canine (and by extension human) body, assessing what constituted a lethal dose and highlighting the symptoms of different forms of poisoning. The British naturalist Charles Waterton, meanwhile, injected the paralytic poison, curare (used by Amazonian peoples for hunting) into a chicken, a hog, a sloth and an ox, recording the dosage required to render each animal unconscious.[70] By the twentieth century fears of malicious poisonings had subsided (in part due to better regulation of substances like arsenic), but toxicity studies continued to play an important role in testing chemical products such as pesticides and cosmetics and ensuring that they were safe for human use. A growing medical establishment also began to dig deeper into the causes of human diseases, using animals as (often literal) guinea pigs in the study of cancer and other serious illnesses. In 1965, for instance, scientists at Louisiana State University taught 'Baboon 007' (nicknamed James Bond) to smoke sixteen cigarettes a day as part of a study into the connection between smoking and lung disease.[71]

A third area in which animal testing has been widely employed is in the development of vaccines. Pioneered in the 1790s, when English physician Edward Jenner proved that exposure to the bovine disease, cowpox, could confer immunity to the much more deadly smallpox, vaccination worked by using less virulent strains of a bacterium or virus to stimulate the production of antibodies in the human (or animal) body. Since Jenner's discovery, thousands of animals have been used in the development of vaccines, serving both as vehicles for reducing the potency of a pathogen (achieved by passing it between individuals until its virulence decreased) and as reservoirs for the disease during its study (virologists could not grow viruses outside of a living medium until the 1930s). The French chemist Louis Pasteur, for instance, conducted experiments on hundreds of rabbits, dogs and (unusually for the time) monkeys in the 1870s and 1880s to develop a vaccine for much-feared cross-species disease, rabies. The American researchers Jonas Salk and Albert Sabin used thousands of rhesus macaques (imported from India) during the development of rival vaccines for the childhood killer polio in the 1950s.[72] While rodents and primates have featured most heavily in vaccine development, other species have also played a role in developing and testing certain vaccines, based on their ability to contract and harbour a particular disease; in the 1970s, for instance, researchers enlisted armadillos in research for a leprosy vaccine because their low body temperatures allowed them to incubate the disease.[73]

Fourth, of course, animal testing has played a crucial role in curing human (and animal) diseases, whether through the development of new drugs and surgical techniques or (less commonly) through the interspecies transfer of bodily parts. From the early twentieth century animals (in this case mice and rabbits) have been used in the development of antibiotics such as Salvarsan (a treatment for syphilis) and penicillin, saving many human lives. From an even earlier period, humans have attempted to use animals as direct donors for failing human organs, using their blood and body parts in a series of cross-species transplants (xenotransplantation). In 1667, for instance, a French physician named Jean-Baptiste Denis attempted inter-animal trans-fusions between dogs, cows, sheep and horses before transfusing sheep's blood into a human patient. To date, the majority of animal-to-human organ transplants have been unsuccessful, resulting in the death (usually within days) of the humans involved (and of course that of the animals that supplied the organs). Experiments are ongoing, however, and scientists are currently attempting to rear genetically engineered pigs whose organs might be implanted in human bodies without inducing rejection.

Finally, in addition to their contributions to anatomy, surgery and the treatment of disease, animals have been used in a wide range of other experi-ments, connected to different areas of human life. In 1889 a group of US scientists subjected multiple animals in Barnum's Menagerie to electric shocks in order to assess the possibility of replacing hanging with electrocu-tion as a form of capital punishment (according to contemporary reports, the wolf 'cried piteously', the seal 'snapped viciously in every direction' and the elephant 'actually enjoyed the sensation in every instance, except when a strong current was applied to the trunk').[74] In the 1950s the American psy-chologist Harry Harlow carried out a series of highly controversial psycho-logical tests on monkeys to explore the impact of maternal deprivation, removing infant monkeys from their mothers and isolating them in a wedge-shaped chamber dubbed the 'pit of despair'. In the 1980s biologist Jo Ann Cameron amputated the limbs of Mexican axolotls (a species of salamander endemic to Lake Xochimilco) to learn more about wound healing and tissue formation (axolotls possess the unique capacity among vertebrates of being able to regenerated an amputated limb).[75] From cosmetic testing to psychol-ogy, therefore, animals have played an important part in the development of new products, medicines and technologies – most designed to prolong or enhance human life, and a few intended to end it.

The Modern Wrack?

While animal experimentation has been central to many aspects of human life it has also become a highly contentious area in human animal relations, raising many complex ethical questions. Are animals the right subjects for testing human medicine? If animals are anatomically similar enough to humans to produce accurate test results, is it ethical to experiment on them? Are some forms of experimentation more acceptable than others, and, if so, why? Have non-human animals benefited from experimentation? For opponents of animal testing, the costs of the process have outweighed the benefits, leading to a range of interrelated criticisms. These criticisms can broadly be divided into two broad areas: concerns about animal suffering and concerns about the impact of vivisection on human society.

Welfare-centred arguments against vivisection have focused primarily on the issue of pain. Animals were, of course, subjected to pain in many walks of life, whether through cruel farming methods, hunting for commerce or sport or overwork. Animal experimentation, however, was seen by some as an exceptionally cruel form of exploitation because it often involved pro-longed and repeated pain, deliberately inflicted in the human quest for knowledge. This elevated it to the level of torture in the minds of anti-vivisectionists, rendering it worse than the suffering experienced by animals slaughtered for food or sport. Writing in the 1880s, for instance, the British antivivisectionist and feminist Frances Power Cobbe reflected:

> The cattle we use for food exist on the condition that we shall take their lives when we need them; and in doing so in the ordinary, not unmerci-ful manner, we save them from the far worse miseries of old age and starvation. To end a creature's existence is one thing. To cause it to suffer torture which shall make that existence a curse is quite another matter.[76]

While the use of anaesthetics from the late nineteenth century served to miti-gate some of the suffering of vivisected animals (and was officially man-dated in Britain in the 1876 Cruelty to Animals Act), this did not eliminate the problem of pain, since certain experiments (for example neurological studies) required subjects to be conscious so that scientists could monitor their reactions. The conditions in which experimental animals were kept, moreover, raised further concerns, especially in the case of intelligent, socia-ble creatures such as primates. In an article from 1982, for example, one journalist highlighted the plight of an eleven-year-old chimpanzee named Larry, who had spent nearly all his life in 'a 3-foot-square cubicle' and had resorted to pulling the hair from his own arms, rocking from side to side and chewing his fingers and toes to cope with the stress of confinement.[77]

Alongside the issue of pain, species has been another key factor in driving opposition to vivisection, with the level of protest very much dependent on the animals involved in a particular experiment. In the nineteenth century, a major reason why many decried vivisection in Europe and North America was because its most common victims were dogs and cats, animals that were often kept as pets. This prompted (often well-founded) fears that companion animals were being snatched from the streets to go under the vivisector's knife. It also elicited emotive allegations of betrayal – especially in the case of dogs. As Cobbe remarked:

By some strange and sinister fatality the chosen victims at present are the most intelligent and friendly of our domestic favourites – the cats who purr in love and confidence as they sit beside us on the hearth; the dogs whose faithful hearts glow with an affection for us truer and fonder than we may easily find in any human breast. To disregard all the beautiful and noble qualities which such animals exhibit, and coldly contemplate them as if their quivering frames were mere machines of bone and tissue which it might be interesting and profitable to explore with forceps and scalpel, is to display heinous indifference to Love and Fidelity themselves, and surely to renounce our claim to be the object of such sentiments to brute or man.[78]

In the twentieth century, the increasing use of primates in scientific research has triggered further qualms from an animal welfare perspective, this time on the basis of the animals' physical and intellectual proximity to humans – a position bolstered by recent zoological and ethological research demonstrating the intelligence of these animals. The excessive demand for rhesus macaques in American research facilities, moreover, has contributed to a severe drop in the macaque population in the animal's native India, where numbers plummeted from an estimated 5–10 million in the 1930s to fewer than 20,000 by the late 1970s.[79] Affective bonds, evolutionary links and environmental concerns have therefore increased public discomfort with using certain species for science, particularly when the direct human benefits of the research are uncertain. By contrast, experiments on rats, mice, guinea pigs, frogs and (since the twentieth century) flies have elicited considerably less opposition, probably because these species are less closely related to human beings.[80]

While anxieties about animal suffering (particularly of primates and companion animals) have thus provided the most fertile ground for opposing animal experimentation, these concerns have been supplemented by other, more anthropocentric worries. First, and particularly prevalent in the nineteenth century, there was a fear that vivisection corrupted the vivisector, creating a moral vacuum at the heart of civilised society. Previously viewed as the preserve of the lower classes, cruelty was now increasingly being committed by middle- and upper-class males, whose professed calling was the preservation of human health. This prompted concerns that cutting up living animals would brutalise practitioners of vivisection, taking away their empathy and humanity – a stereotype propagated in much contemporary antivivisection literature. During the public vivisection of a dog, for instance, witnessed by humanitarian Louise Lind af Hageby, one unnamed experimenter reportedly joked that "'the dog's intestine might fall out" if care is not taken to keep it in place' – a remark that was 'fully appreciated by the audience [made up of medical students], who applaud[ed] and laugh[ed]'.[81]

Second, and closely linked to anxieties about the impact of vivisection on the vivisector, there was a widespread (and in many cases justified) concern that experimenting on animals would soon extend into experimentation on other subaltern groups – from colonial peoples to humans with disabilities – the 'slippery slope' argument. To cite Cobbe once again, 'If it be proper to torture a hundred affectionate dogs or intelligent chimpanzees to settle some curious problem about their brains, will they advocate doing the same to a score of Bosjesmen [a contemporary name for the San people of southern Africa], to the idiots in our asylums, to criminals, to infants, to women?'[82] Such concerns were, of course, borne out by subsequent experience, most shockingly under the Nazis, who experimented on humans in concentration camps. They also raised wider issues about informed consent in medical trials, some of which – even outside of Nazi Germany – adopted practices that we would now view as unacceptable. Salk, for instance, tested his polio vaccine on developmentally delayed children in a Pennsylvania school after initial testing on monkeys, something that would be considered unethical today.[83]

Lastly, and perhaps most fundamental to the whole business of animal testing, some opponents have raised the more prosaic question of whether it constitutes a valid means of assessing the human body – in short, can experimentation on animals produce reliable results for human medicine? In some cases, of course, animal bodies have offered useful substitutes for human ones when the latter were unavailable, and some medical breakthroughs could not have been made without them. Successful vaccines for rabies, polio and hepatitis, for instance, all relied on animal testing during their development. In other instances, however, it is arguable that animal experimentation – at least when done on the wrong species – has caused errors and delayed accurate understandings of the human body, and even, in some cases, resulted in active harm to human patients. Galen's dissection of barbary apes to study the circulation, for instance, led him to believe that blood flowed through holes in the septum of the heart, when in fact (for humans) this is not the case; this belief endured until the sixteenth century. In 1957, meanwhile, scientists promoted the drug Thalidomide as an anti-nauseant for pregnant women, having first tested it on rats and rabbits with no ill-effects. Thalidomide subsequently caused severe birth defects in the babies of the women who had taken it (Thalidomide also causes birth defects in primates, but they were not initially used as test subjects).[84] Animal experimentation could therefore lead to incorrect conclusions if the animal bodies selected did not accurately mirror the human bodies on which the resulting medicine would be used. This, however, returns us to one of the key paradoxes of animal experimentation, and a central point in the vivisection debate. To be useful as an experimental subject, an animal must be close enough to humans to act as a proxy for the human body. But if a species is sufficiently close to humans to replicate their physiological functions, is it not also close enough to feel pain in the same way as we do, and, if so, is it ethically acceptable to use it as a test subject?

The Baboon Seven

To explore some of the complex ethical issues surrounding animal testing, we conclude by looking at a particularly controversial case from the 1970s: the use of live baboons as crash test dummies. Aimed at improving road safety through the development of seat belts and air bags, this experiment was justified by its funders, the Ford Motor Company, as a way of saving human lives. Opponents of the practice, however, contended that it was both cruel and unnecessary. They lobbied vocally for its termination.

The use of baboons in simulated crashes began in 1967 and was designed to assess the effects of high-speed motor vehicle collisions on the human body. In a series of tests conducted, initially, at Holloman Airforce Base, New Mexico, and, later, at the University of Michigan, Ann Arbor, live baboons were strapped onto sleds and propelled at speed into an object that inflicted severe chest injuries. The primates were anaesthetised before the experiments and killed before they could regain consciousness. The tests were repeated at different speeds, with and without seat belts and, subsequently, with air bags, in order to determine the effectiveness of the latter. It was found that 'when ploughing into the air bag at impact, the baboons survived the equivalent of crashes up to 64 miles an hour, while others were killed at far lower speeds while wearing other restraint systems employing seat belts'.[85]

While motor industry executives considered it acceptable to sacrifice baboons to improve human safety, others took a different view, arguing that it was wrong to use primates for this purpose. These criticisms took several forms. First, even though the baboons were unconscious at the point of impact, animal rights campaigners claimed that the experiments constituted an 'appalling abuse of animals'.[86] Second, elaborating on this view, other opponents pointed to the environmental impact of extracting baboons (and other primates) from the wild for use in experiments, which, in their opinion, contravened contemporary efforts to conserve endangered species. As one critic, the Reverend Erwin Grede, expressed it: 'It is frightening and appalling to think that people would treat animals in this way, particularly at a time of rising consciousness and concern over conservation; it flies in the face of decency and concern for living things'.[87] Third, some critics pointed out the potential limitations of using baboons as proxies for humans, noting that physiological differences between the two species made the results of the crash tests unreliable. To quote one writer: 'They are the wrong size and shape, and merely because a test crash ruptures a monkey's [sic] spleen or lacerates its aorta, no proof has been established that a human will suffer the same injury'.[88] Baboon crash tests were therefore abusive to the individual animals, a threat to the survival of the species and potentially pointless anyway because they could not properly replicate what would happen to a human being.

Responding to these criticisms, scientists and motor companies defended the validity of the tests, insisting that the benefits outweighed the costs. Baboons, they claimed, provided a good model for human crash victims and the animals used did not suffer pain, as they were unconscious. While other ways of gathering data existed, moreover, these also came with ethical challenges. One option, for instance, the replacement of living baboons with human corpses, donated for medical research, elicited opposition from religious groups, who considered it wrong to use the bodies of human beings in this way. Another option, the use of living human beings, proved a nonstarter, given the absence of volunteers – though one journalist suggested that motorists convicted of speeding might be offered the option of participating in collision tests instead of getting points on their licence, given that 'on the basis of the way they drove, they wouldn't live much longer anyway'.[89] A third option, the use of bespoke crash test dummies specially designed to mimic the human body, proved much more acceptable. In order to manufacture these dummies accurately, however, scientists claimed that it was first necessary to perform similar tests on animals, thus perpetuating the controversial experiments. Ultimately, the University of Michigan ended its baboon crash tests in 1978 and the surviving six members of the so-called 'baboon seven' were reprieved. Though an apparent victory for humanitarians, this decision seems, in fact, to have reflected the priorities of the scientists, who had by then collected all the data they required.

The use of baboons for over a decade in road safety tests illustrates some of the key dilemmas posed by the use of animals in research – and the factors that shaped its reception. First, the level of perceived suffering experienced by the test subject was important, as this, for many people, determined the heinousness of the experiment. The baboon crash tests were arguably at the more humane end of the experimental spectrum, given that the baboons in question were anaesthetised during the tests and euthanised immediately afterwards. For some, however, the fact that the animals were subjected to the tests at all was an abomination. Second, species was of critical importance in determining how animal research was received, and, in this case, appears to have played a significant role. Rats, mice, frogs or even pigs might have generated less outrage, but baboons, closely related to humans, were seen as being more intelligent and having greater capacity for suffering. Experiments on dogs or chimpanzees would likely have elicited similar opposition. Lastly, the purpose of the experiment also influenced how people responded to animal testing, with greater anger directed towards those experiments that offered fewer potential benefits to humans or were seen as gratuitous. In the case of the crash-test baboons, the benefits were potentially great, but there was also a perception among critics that road safety research was less worthy than testing to further medical knowledge, perhaps in part because some of those humans killed or injured on the road were seen to have brought about their own misfortune. As one journalist commented:

> Most animal baboons I talked to … [said that] they don't mind going up in space, and they're even willing to participate in cancer experiments. But they feel they shouldn't be wasted in auto safety tests, particularly since there are so many human baboons available.[90]

This point is, of course debatable, but it is noteworthy that contemporary medical experiments on baboons generated less potent opposition – at least from an animal welfare perspective. It is also noteworthy that the six baboons at Ann Arbor were, following their reprieve from crash tests, transferred to the school's physiology department for use in a study of hypertension – a move that elicited few objections.[91] Using baboons to help curb high blood pressure, treat cancer or provide humans with replacement hearts or livers was thus seemingly more acceptable than using them to limit the impact of motor vehicle accidents.

Conclusion

Humans have always been curious about other animals and wanted to learn more about them. From paintings of bison in the caves at Altamira to Attilio Gatti's first blurry photograph of a wild okapi, humans have sought to capture animal forms for posterity, documenting their movements and behaviours. Across different cultures, humans have also sought to classify other species, using their anatomies, habits, diets and reproductive strategies to place them in different categories – bird/mammal, carnivore/herbivore, wild/ domesticated. Underlying much research into the physiology, psychology and cognitive capacity of animals has been the deeper question of what, exactly, separates non-humans from humans – brain size, tool use, language, culture? – definitions that have been repeatedly challenged by tool-using chimpanzees, signing gorillas and elephants apparently able to recognise their reflections in a mirror.

In addition to studying animals as specimens of natural history, humans have experimented on other species for their own benefit, using non-human animals as proxies for the human body. Experiments on animals have contributed to advances in human (and veterinary) medicine, helping to improve surgical procedures, develop new drugs, predict the likely course of diseases and – in the case of the baboon seven – improve humans' chances of surviving a road traffic collision. Like other uses of animals, however, vivisection has generated intense opposition, prompting questions about the ethics and effectiveness of using guinea pigs, dogs or monkeys as human surrogates. Are non-humans suitable for testing medicines intended for humans? Should some species be excluded from experimentation for welfare reasons? Might testing on animals spill over into non-consensual testing on vulnerable human beings? As studies in ethology advance our understanding of animal sentience and cognition, these questions become ever more pressing. If dogs can express human-like emotions, capuchin monkeys can use tools and sheep can recognise faces, is it right to transplant their organs, send them into space or use their bodies in vehicle crash tests?

Notes

1 Thomas Nagel, 'What is it Like to Be a Bat?', *The Philosophical Review* 83:4 (1974), pp. 435–450.
2 *The Luttrell Psalter*, England (Lincolnshire), *c*.1320–1340, BL Add. MS 42130, f.169v, f.59v and f161r.
3 Karen Meir Reeds and Tomomi Kinukawa, 'Medieval Natural History', in David Lindberg and Michael Shank (eds), *The Cambridge History of Science* (Cambridge, Cambridge University Press, 2013), pp. 569–589.
4 Gary Urton, 'Animals and Astronomy in the Quechua Universe', *Proceedings of the American Philosophical Society* 125:2 (1981), pp. 110–127.
5 Frank Egerton, 'A History of the Ecological Sciences, Part 6: Arabic Language Science – Origins and Zoological Writings', *Bulletin of the Ecological Society of America* (2002), pp. 142–145.
6 Clements R. Markham (ed., trans.), *The Travels of Pedro Cieza de León, A.D. 1532–50, Contained in the First Part of his Chronicle of Peru* (London, 1864), vol. I, p. 393.
7 Miguel Asúa and Roger French, *A New World of Animals: Early Modern Europeans on the Creatures of Iberoamerica* (Aldershot: Ashgate, 2005).
8 Edward Topsell, *The History of Four-Footed Beasts, Serpents and Insects* (London: E. Cotes, 1658), pp. 28–34.
9 Harriet Ritvo, *The Platypus and the Mermaid and Other Figments of the Classifying Imagination* (Cambridge, MA: Harvard University Press, 1997), pp. 1–50.
10 'Chimpanzees Seen Making Tools', *The Times*, 28 February 1964.
11 'Sheep Faces Leave Impression', *Washington Post*, 12 November 2001.
12 'Chimpanzees Seen Making Tools', *The Times*, 28 February 1964.
13 'When the Gorilla Speaks', *Washington Post*, 31 January 1985.
14 On animal cognition and the history of ethology see Frans de Waal, *Are We Smart Enough to Know How Smart Animals Are?* (London: Granta, 2016).
15 On the challenges posed by field studies see Stuart McCook, '"It may be truth, but it is not evidence": Paul du Chaillu and the legitimation of evidence in the field', *Osiris* (2nd series) 11 (1996), pp. 177–197; and Amanda Rees, *The Infanticide Controversy: Primatology and the Art of Field Science* (Chicago: University of Chicago Press, 2009).

16 On the emergence of field stations see Megan Raby, *American Tropics: The Caribbean Roots of Biodiversity Science* (Chapel Hill: University of North Carolina Press, 2017).

17 'The Zoological Gardens', *The Times*, 11 October 1909.

18 'Monkey Think, Monkey Do', *Washington Post*, 12 April 1995.

19 'Who's That Pretty Pachyderm', *Washington Post*, 31 October 2006.

20 Iris Montero Sobrevilla, 'Indigenous Naturalists', in Helen Curry, Emma Spary, James Secord, and Nick Jardine (eds), *Worlds of Natural History* (Cambridge: Cambridge University Press, 2018), pp.112–130'.

21 Jane Camerini, 'Wallace in the Field', *Osiris* (2nd series) 11 (1996), pp. 44–65.

22 Nancy Jacobs, *Birders of Africa: History of a Network* (New Haven: Yale University Press, 2016), pp. 148–220.

23 'The Allure of Elephants', *New York Times*, 10 March 1988.

24 'Chimpanzees Seen Making Tools', *The Times*, 28 February 1964.

25 Rachel Poliquin, *The Breathless Zoo: Taxidermy and Cultures of Longing* (University Park: Pennsylvania State University Press, 2012), pp. 11–42.

26 Mary Anne Andrei, *Nature's Mirror: How Taxidermists Shaped America's Natural History Museums and Saved Endangered Species* (Chicago: University of Chicago Press, 2020), pp. 181–190.

27 Charlotte Sleigh, *The Paper Zoo: 500 Years of Animals in Art* (London: The British Library, 2016), pp. 26–27; Gregg Mitman, *Reel Nature: America's Romance with Wildlife on Film* (Seattle: University of Washington Press, 1999).

28 Etienne Benson, *Wired Wilderness: Technologies of Tracking and the Making of Modern Wildlife* (Baltimore: Johns Hopkins University Press, 2010).

29 'Rhinos in South Africa are Bugged for Science', *New York Times*, 15 August 1970.

30 'New Research on Malaysia's Odd, Elusive Tapirs', *New York Times*, 2 June 2009.

31 'Hunted Gorillas', *The Times*, 10 June 1924.

32 'Grizzlies Bugged for Science's Sake', *New York Times*, 7 August 1963.

33 'Do Two Research Chimps Want to Retire?', *New York Times*, 10 June 1982.

34 Bernabe Cobo, *Historia del Nuevo Mundo* (Sevilla: Imprenta de E. Rasco, 1891), vol. II, pp. 146–147.

35 'Science Jottings: Manatee', *Illustrated London News*, 11 May 1889.

36 For a discussion of the classificatory disputes surrounding another liminal animal, the whale, see D. Graham Burnett, *Trying Leviathan: The Nineteenth-Century Court Case that Put the Whale on Trial and Challenged the Order of Nature* (Princeton: Princeton University Press, 2007).

37 James Murie, 'On the Form and Structure of the Manatee (*Manatus americanus*), *Transactions of the Zoological Society of London* Vol. VIII (1874), pp. 127–202.

38 Agnes Crane, 'Notes on the Habits of the Manatees (*Manatus australis*) in Captivity in the Brighton Aquarium', *Proceedings of the Zoological Society of London* (London: Printed for the Society, 1881), p. 460.

39 'Piel de Maniatí', Archivo General de Indias, Indiferente 1549.

40 'Manatee V, the New Sea Cow', *New York Times*, 28 October 1906.

41 'Manatee's Odyssey Now is no Oddity', *Washington Post*, 4 July 1996.

42 Roger Reep and Robert Bonde, *The Florida Manatee: Biology and Conservation* (Gainesville: University of Florida Press, 2021), pp. 126–156.

43 'Manatee V, the New Sea Cow', *New York Tribune*, 28 October 1906.

44 George Shaw and Frederick Nodder, 'The Duck-Billed Platypus', *The Naturalist's Miscellany*, Volume 10, 1799 (no page numbers).

45 Everard Home, 'A Description of the Anatomy of the Ornithorhynchus paradoxus', *Philosophical Transactions*, 1 January 1802, pp. 67–84.
46 Richard Owen, 'On the Mammary Glands of the *Ornithorhynchus paradoxus*', *Philosophical Transactions*, 1832, pp. 517–534.
47 'Australian Fauna', *Pall Mall Gazette*, 20 January 1897.
48 'Platypus Born in Captivity', *The Times*, 5 January 1944.
49 'Platypuses Mark 10th Year at Zoo', *New York Times*, 30 April 1957.
50 'Painkiller Hope in Platypus Poison', *The Times*, 1 January 1998.
51 'Looking at Genome of the Platypus', *New York Times*, 8 May 2008.
52 'Australian Fauna', *Pall Mall Gazette*, 20 January 1897.
53 'Platypuses to Come Out to Play', *New York Times*, 30 April 1947.
54 Robert Paddle, *The Last Tasmanian Tiger: The History and Extinction of the Thylacine* (Cambridge: Cambridge University Press, 2000), pp. 7–8. For a more recent exploration of this topic, see also Jack Ashby, *Platypus Matters: The Extraordinary Story of Australian Mammals* (London: HarperCollins, 2022).
55 'Sir Harry Johnston's Recent Journeys in the Uganda Protectorate', *The Times*, 29 December 1900; 'A New Mammal', *The Times*, 7 May 1901.
56 'Sir Harry Johnston's New Beast', *The Times*, 18 June 1901.
57 'The Okapi', *The Times*, 19 October 1901.
58 'The Okapi', *The Times*, 19 October 1901.
59 'Okapi', *The Times*, 27 October 1906.
60 Attilio Gatti, *Great Mother Forest* (London: Hodder and Stoughton, 1936), pp. 202–284.
61 'Animal Reluctantly Enters Spotlight', *Chicago Tribune*, 13 November 1997.
62 'A Live Okapi in Europe', *The Times*, 15 August 1919.
63 'Rare Okapi Here for Zoo in Bronx', *New York Times*, 3 August 1937.
64 Gatti, *Great Mother Forest*, pp. 290, 226 and 85.
65 'Congo Nurtures Okapi for Export', *New York Times*, 29 September 1963.
66 'Rare Chimpanzee at the Zoo', *The Times*, 18 September 1937.
67 'The Okapi', *Chicago Daily Tribune*, 7 January 1951.
68 'Professor Ray Lankester on the Okapi', *The Times* 8 February 1902; 'Okapi', *The Times*, 27 October 1906.
69 'The Okapi', *The Times*, 8 May 1936.
70 Charles Waterton, *Wanderings in South America* (London: Cassell, 1891), pp. 86–110.
71 'Scientists Teach Baboon to Smoke for Cancer Study', *New York Times*, 20 June 1965.
72 Anita Guerrini, *Experimenting with Humans and Animals* (Baltimore: Johns Hopkins University Press, 2022), pp. 102–144.
73 'Medicine: Studies of Leprosy', *The Times*, 8 April 1974.
74 'It Tickled the Elephants', *The Evening Telegraph and Star*, 16 February 1889.
75 'On a Limb: Scientists Cut and Slash Amphibians Seeking Clues to the Regeneration of Tissue', *Chicago Tribune*, 19 December 1985. On the wider use of axolotls in science, see Christian Reiss, Lennart Olsson and Uwe Hossfeld, 'The History of the Oldest Self-Sustaining Laboratory Animal: 150 Years of Axolotl Research', *Journal of Experimental Zoology* 32B (2015), pp. 393–404.
76 Frances Power Cobbe, 'The Moral Aspects of Vivisection', *The Modern Rack* (London, 1889), p. 10.
77 'Experiment Breeds Life into Chimps' Future', *Chicago Tribune*, 22 July 1982.
78 Power Cobbe, 'The Moral Aspects of Vivisection', p. 15.
79 Guerrini, *Experimenting with Humans and Animals*, p. 134.

80 Paul White, 'The Experimental Animal in Victorian Britain', in Lorraine Daston and Gregg Mitman (eds), *Thinking with Animals: New Perspectives on Anthropomorphism* (New York: Columbia University Press, 2005), pp. 59–82; Joanna Dean, 'Guinea Pig Agnotology', in Jennifer Bonnell and Sean Kheraj (eds), *Traces of the Animal Past: Methodological Challenges in Animal History* (Calgary: University of Calgary Press, 2021), pp. 175–197.
81 Louise Lind-Af-Hageby and Leisa Katherina Schartau, *The Shambles of Science: Extracts from the Diary of Two Students of Physiology* (London: Earnest Bell, 1903), pp. 139–148.
82 Power Cobbe, 'The Moral Aspects of Vivisection', p. 7.
83 Guerrini, *Experimenting with Humans and Animals*, p. 140.
84 Guerrini, *Experimenting with Humans and Animals*, p. 150.
85 '9 Baboons Test Air Bag Restraint', *Washington Post*, 26 October 1967.
86 'Ford to Go on Using Baboons for Tests', *New York Times*, 19 December 1967.
87 'Outcry in Michigan over Seven Baboons Doomed in Car Crash Experiments', *The Times*, 6 February 1978.
88 'Detroit's Deathless Daredevils', *Chicago Tribune*, 31 March 1968.
89 'A Baboon Boom Boom', *Washington Post*, 26 November 1967.
90 'A Baboon Boom Boom', *Washington Post*, 26 November 1967.
91 '5 Baboons, One Reprieved, Face Death in Second Lab', *New York Times*, 13 March 1978.

Further Reading

Asúa, Miguel, and Roger French, *A New World of Animals: Early Modern Europeans on the Creatures of Iberoamerica* (Aldershot: Ashgate, 2005)

Benson, Etienne, *Wired Wilderness: Technologies of Tracking and the Making of Modern Wildlife* (Baltimore: Johns Hopkins University Press, 2010)

Burnett, D. Graham, *Trying Leviathan: The Nineteenth-Century Court Case that Put the Whale on Trial and Challenged the Order of Nature* (Princeton: Princeton University Press, 2007)

Camerini, Jane, 'Wallace in the Field', *Osiris* (2nd Series) 11 (1996), pp. 44–65

Dean, Joanna, 'Guinea Pig Agnotology', in Jennifer Bonnell and Sean Kheraj (eds), *Traces of the Animal Past: Methodological Challenges in Animal History* (Calgary: University of Calgary Press, 2021), pp. 175–197

De Waal, Frans, *Are We Smart Enough to Know How Smart Animals Are?* (London: Granta, 2016)

Guerrini, Anita, *Experimenting with Humans and Animals* (Baltimore: Johns Hopkins University Press, 2022)

Jacobs, Nancy, *Birders of Africa: History of a Network* (New Haven: Yale University Press, 2016)

McCook, Stuart, '"It May Be Truth, but it Is Not Evidence". Paul du Chaillu and the legitimation of evidence in the field', *Osiris* (2nd Series) 11 (1996), pp. 177–197

Mitman, Gregg, *Reel Nature: America's Romance with Wildlife on Film* (Seattle: University of Washington Press, 1999)

Montero Sobrevilla, Iris, 'Indigenous Naturalists', in Helen Curry, Emma Spary, James Secord, and Nick Jardine (eds), *Worlds of Natural History* (Cambridge: Cambridge University Press, 2018), pp.112–130

Nagel, Thomas, 'What is it Like to Be a Bat?', *The Philosophical Review* 83:4 (1974), pp. 435–450

Poliquin, Rachel, *The Breathless Zoo: Taxidermy and Cultures of Longing* (University Park: Pennsylvania State University Press, 2012)

Raby, Megan, *American Tropics: The Caribbean Roots of Biodiversity Science* (Chapel Hill: University of North Carolina Press, 2017)

Rees, Amanda, *The Infanticide Controversy: Primatology and the Art of Field Science* (Chicago: University of Chicago Press, 2009)

Reiss, Christian, Lennert Olsson and Uwe Hossfeld, 'The History of the Oldest Self-Sustaining Laboratory Animal: 150 Years of Axolotl Research', *Journal of Experimental Zoology* 32B (2015), pp. 393–404.

Ritvo, Harriet, *The Platypus and the Mermaid and Other Figments of the Classifying Imagination* (Cambridge, MA: Harvard University Press, 1997)

Sleigh, Charlotte, *The Paper Zoo: 500 Years of Animals in Art* (London: The British Library, 2016)

Urton, Gary, 'Animals and Astronomy in the Quechua Universe', *Proceedings of the American Philosophical Society* 125:2 (1981), pp. 110–127

White, Paul, 'The Experimental Animal in Victorian Britain', in Daston, Lorraine, and Mitman, Gregg (eds), *Thinking with Animals: New Perspectives on Anthropomorphism* (New York: Columbia University Press, 2005), pp. 59–82

6   Biodiversity

In 1768 the last Steller's sea cow was hunted and killed in the Bering Sea. A type of gigantic manatee, reaching up to thirty feet in length, the sea cow had once ranged as far south as Baja California and as far west as the shores of the Kamchatka Peninsula in Russia. Since its discovery by Russian explorers in 1741, however, the species' population had plummeted dramatically, the direct result of hunting by fur traders, who used the animal's flesh for food.[1]

Two centuries later, the sea cow's close relative, the manatee, also came under significant pressure from humans, in this case primarily as a result of habitat loss, pollution and injuries from collisions with boats in the Florida Everglades. On this occasion, however, rather than accepting the animal's inevitable demise, humans exerted themselves to save the species. In 1893 the state of Florida introduced a law prohibiting manatee hunting. Beginning in the 1980s, further, more comprehensive, efforts were made to preserve the manatee, from the introduction of speed limits for motor boats to the rescue and treatment of sick or injured animals in special marine sanctuaries.[2] In 1986 the US Fish and Wildlife Service rescued and rehabilitated a manatee called Adair, 'when she nearly drowned after becoming entangled in a crab-trap line'.[3] In 1999 a team of veterinarians and neurosurgeons performed a five-hour spinal surgery on a manatee named Nash after a 'boat's propeller sliced away 25 pounds (11kg) of skin away from [his] backbone'.[4] Though largely responsible for the manatee's decline, therefore, humans are also going to great efforts to reverse it, in part due to the species' iconic status as one of Florida's most recognisable native mammals. In other ways, moreover, manatees have successfully adapted to close cohabitation with humans, feeding on fast-growing freshwater weeds introduced by humans and (since the 1950s) congregating around the outflows of power plants to stay warm during winter – though in recent years algal blooms induced by sewage and fertiliser runoff have severely diminished the species' food supply.

The differing fates of the sea cow and the manatee illustrate the multiple ways in which human activities have impacted on other species. By cutting

DOI: 10.4324/9781003181996-7

down forests, hunting animals for their meat and skins, polluting the seas and rivers and moving plants, animals and microbes between continents, humans have threatened the survival of other species, pushing some to extinction. At the same time, some humans in some periods and places have gone to extreme lengths to protect some species of animals from obliteration, demonstrating a powerful urge to keep certain valued animals in existence – whether to enable their continued exploitation or for less utilitarian reasons. Examining both trends, this chapter explores the impact of humans on global biodiversity and assesses the extent to which humans have been responsible for the diminution or increase of animal species. How has the rate of animal extinction changed over time? Have all humans been equally responsible for the decline in other animal populations, or are there temporal and geographic differences? To what extent have conservation efforts proven effective? What practical and ethical challenges have they presented?

Biodiversity

Humans have had a largely negative impact on global biodiversity (a term that first became widespread in the 1980s). Some species, of course, have disproportionately increased over the centuries due to human intervention. Domesticated animals, raised for meat and milk, have reproduced in far higher numbers than would have been possible in the wild, thanks to artificial feeding and predator control.[5] Versatile scavengers such as rats and mice have also thrived in close proximity to humans. For most species, however, living alongside humans has resulted in population decline, in some cases to the point of extinction. This process began in the Pleistocene era, when human hunting probably annihilated species such as the giant ground sloth, but has accelerated since the Columbian Exchange and the Industrial Revolution. It has become particularly intense since the so-called Great Acceleration of the 1950s, when global (human) population growth, intensive agriculture and deforestation happened at an unprecedented rate. The World Wildlife Foundation estimates that there has been an average 69% decline in the relative abundance of monitored wildlife populations around the world between 1970 and 2018, with between 0.01% and 0.1% of species currently becoming extinct each year – an extinction rate between 1,000 and 10,000 times higher than the natural extinction rate (i.e. the species that would have gone extinct anyway without human interference).[6]

What have been the causes of this acceleration in biodiversity loss? Biologists point to five key factors, which can be summarised under the acronym HIPPO: habitat destruction, invasive species, pollution, human population growth and overharvesting.[7] Let's look at each of these in turn.

Habitat loss has been (and remains) a major driver of biodiversity loss across the globe. Human encroachment into animal habitats dates back from the very earliest period of human settlement, when humans felled trees for timber and cleared land for pasture and crop cultivation. In Central America, for example, the Maya (*c*.1800BC–900AD) cut down almost all the native rainforest and replaced it with crops, such as maize, leading to soil erosion and drought.[8] Since the sixteenth century – and especially since the nineteenth century – this process has intensified, primarily due to the expansion of cash crops such as sugar, bananas and palm oil and the growing global demand for meat (see Chapter 1). In the Caribbean, coastal mahogany forests were razed in the sixteenth century to make way for sugar plantations. In Indonesia, huge swathes of rainforest have been cleared since the 1970s for the cultivation of oil palms (native to West Africa but introduced into East Asia in the early twentieth century), decimating populations of Sumatran rhinoceroses, Sumatran tigers and orangutans. Habitat fragmentation, meanwhile, has left some animal populations isolated and vulnerable to inbreeding, as roads, canals and fences have obstructed longstanding migration routes. In 2016, for instance, President Donald Trump's proposed border wall between the USA and Mexico generated vocal opposition from conservationists, who warned that it would prevent migratory animals like wolves, black bears and – most famously – a male jaguar nicknamed 'El Jefe' (the only one of his species then known to have entered the USA in recent decades) from moving between the two countries.[9]

Invasive species have likewise served as a major driver of biodiversity loss, though their effects have been felt more intensively in some places than in others. Animals have, of course, always moved between landmasses to some degree, crossing land bridges during periods of global cooling (and correspondingly lower sea levels), or, in the case of birds and marine species, flying or swimming to new habitats. Human technologies, however, in the form of ships, and later planes, have greatly accelerated this process, bringing many species to lands that they would not have reached without human intervention; in the early 1900s, for instance, over twenty murine opossums reached New York from South America 'in bunches of bananas'.[10] While the majority of introductions (both deliberate and accidental) have failed, leaving little mark on the historical record, a few introduced species have prospered in their new environments, often with devastating consequences for local ecosystems and native species. Following the Spanish conquest of Mexico, for example, populations of introduced cows and sheep surged on the colony's northern grasslands, overgrazing the native flora and permanently changing the landscape.[11] In seventeenth-century Mauritius, the introduction of pigs and macaques by Dutch sailors hastened the demise of the dodo – already under severe pressure from human predation.[12] In Australia, meanwhile, South American cane toads, introduced deliberately to Queensland in 1935 to control the cane beetle, have proven lethal to native species like the goanna lizard and the northern quoll, which can perish from just one lick of their highly toxic skin.[13] While humans have thus assisted in the multiplication of some species – such as cats, rats and cows – their transcontinental movements have contributed directly to the decline of others by bringing them into contact with new and dangerous biota.

A third contributor to biodiversity loss has been pollution. Oil slicks, discarded fishing tackle and plastic waste have wrought devastation among fish, birds and marine mammals. Pesticides like DDT have had a catastrophic impact on wildlife, while light and sound pollution have disrupted animal migration patterns, possibly contributing to mass whale strandings. Many animal deaths have also been brought about by road and maritime traffic since the early twentieth century, as cars, trains and boats extend to new areas of the globe. A survey in 2002 found that 97% of the 1,250 adult manatees in Florida's state database had 'more than one set of propeller and keel scars', while one animal had been struck 50 times – the last of these fatally.[14] Human-caused pollution is not an exclusively modern phenomenon, and it is possible to identify examples of localised contamination in earlier historical periods. The Spanish silver mines at Potosí, for instance, caused considerable levels of water and soil pollution in the seventeenth century, shortening the lives of the llamas and mules used there as beasts of burden – and also those of the indigenous workforce. While some major pollution blackspots existed before 1800, however, these were usually confined to a comparatively small area and did not threaten the survival of entire species. By contrast, nineteenth-century pollution – and above all twentieth- and twenty-first-century pollution – has affected much larger swathes of the globe, as man-made chemicals, oil and plastics have contaminated the land, air and oceans.

Over-harvesting – to take our factors out of order – has had perhaps the most obvious and direct impact on wild animal populations, and, unlike pollution, extends back over many centuries. Animals have been hunted for sustenance, for sport and for commercial reasons. African grey parrots were (and are) hunted for the exotic pet trade, along with many other wild bird species. Beavers, fur seals and chinchillas were hunted to the point of extinction for their precious fur coats, while pangolins and rhinoceroses are today killed in large numbers for their scales and horns, both valued in 'traditional' Chinese medicine. Species that have lacked direct value for humans, meanwhile, have often been classed as vermin and culled to protect domesticated crops and livestock. Wolves, for example, were exterminated in Hokkaido (the northern-most island of Japan) in the 1880s and 1890s to safeguard the island's horses (bred there in growing numbers following the Meiji Restoration of 1868).[15] Pumas, guanacos and wild dogs were culled in Tierra del Fuego in the early twentieth century to make way for sheep.[16] Though many animal populations have been able to tolerate a certain level of human predation, the extent and intensity of hunting has often rendered it unsustainable, killing animals more quickly than they have been able to reproduce. Commercial hunting, in particular, has been devastating to many species, with the capitalist imperative for profit encouraging excess slaughter. As the British soldier and game warden Henry Courtney Brocklehurst commented in 1939: 'When a market is created for a living animal or bird, or some part of it which entails the taking of life, whether it be the giant panda, the horn of the rhinoceros, the gland of the musk deer or the feathers of the egret, it undoubtedly becomes the thin end of the wedge of extermination'.[17]

As for (human) population growth, this has exacerbated all the factors described above, putting animals in further jeopardy. Relatively stable until around 1800, the global human population began to climb steadily in the eighteenth century and steeply from 1950, rocketing from around 1 billion in 1800 to an estimated 7.7 billion in 2019.[18] All those additional human beings had to be fed, which has meant clearing more land for agriculture. Increasing numbers of humans wanted to eat meat, which has meant more cattle, sheep and pigs, and more deforestation. Where existing territories were unable to sustain growing populations, humans migrated, colonising sparsely populated regions and bringing many non-human animals with them – some of which became invasive species. And the more humans there were on earth, the greater the demand for the skins, bones and teeth of wild animals, and the higher the levels of pollution. While the number of domesticated animals has thus increased in proportion to the rise in the human population, the number of wild animals has decreased, pushing many to the edge of extinction.

Lastly, in addition to HIPPO, we should add one new factor to our explanations of biodiversity loss: human-induced climate change. Since the Industrial Revolution, humans have pumped increasing amounts of carbon dioxide and methane into the atmosphere, raising global temperatures, melting ice caps and glaciers, changing weather patterns and acidifying seas. This has been devastating for many non-human animals, which cannot evolve quickly enough to adapt to such dramatic changes. The poster child for climate change, the polar bear, has seen its Arctic habitat rapidly depleted by rising temperatures, pushing survivors into dangerous conflict with human communities; in 2019, 56 starving polar bears descended on the Russian village of Ryrkaypiy, in the Chukotka region in search of food because the coastal ice was too thin for them to stand on it and hunt seals.[19] Many other species have also been affected by the myriad changes to the world's climate, from the 60,000 koalas estimated to have perished in Australia's devastating bush fires in 2020 to the 200,000 saiga antelope that expired in Kazakhstan in 2015 from haemorrhagic septicaemia, a disease caused by a bacterium (*Pasteurella multocida*) that multiplies more easily in warm, humid conditions.[20] Like population growth, therefore, climate change is exacerbating existing challenges for non-human animals, making their survival increasingly perilous.

Lost Species, Damaged Ecosystems

HIPPO gives historians a useful framework for understanding the general causes of biodiversity loss over the past millennium. To explain specific population declines and extinctions, however, we need to take a more granular approach, and to explore how these different factors operated in conjunction with one another to bring about a collapse. How can we account for the loss of specific species in specific places and periods? How has the decline of particular species impacted on the wider ecosystems of which they formed a part? What implications has their loss had for other animals – human and non-human?

To answer the first question, it is instructive to look in more detail at a couple of case studies: the North American bison and the Tasmanian tiger, both of which succumbed to a range of different pressures – all human-induced. In the case of the bison, hunting was the main issue, combined with habitat loss, human encroachment and disease. Once present in vast numbers on the Great Plains, bison had for millennia been hunted by First Nations peoples on a subsistence basis – a process that became easier, and more intensive, after the introduction of horses from the late seventeenth century. From the mid-nineteenth century the level of exploitation expanded massively, as steamships and railways made the herds more accessible, more powerful rifles facilitated larger kills, and demand increased for bison robes, bison hides (used to make leather belts for textile mills) and bison bones (used in sugar refining and as a fertiliser). Afflicted simultaneously by drought, introduced diseases and competition for forage from Old-World cattle, the bison population was unable to withstand this relentless assault, collapsing from as many as 30 million in the mid-eighteenth century to a few hundred by the early twentieth century. A last-ditch effort from conservationists saved the species from extinction, but the bison who survived in the newly established private reserves and national parks were domesticated captives, not truly wild animals.[21] Today the majority of North America's 340,000 bison live on private farms and ranches, where they are reared to produce bison burgers – a sad end for a species once synonymous with wilderness.[22]

While the bison survived by the skin of its teeth, the Tasmanian tiger succumbed to extinction in the early twentieth century as a result of similar challenges. Native to the Antipodes, the Tasmanian tiger – also known as the thylacine – had once existed across Australia, but, by the nineteenth century, was confined to the southern island of Tasmania. A kind of marsupial wolf, the thylacine survived largely on a diet of small native mammals, such as wallabies and wombats, and had little contact with humans. In the nineteenth century, however, European settlers accused it of killing their sheep and began a sustained campaign of persecution against the predator, placing a bounty on its head from 1888 and rewarding anyone who could produce evidence of a dead tiger. Coupled with habitat loss, introduced diseases, and the decline of its native prey species, unremitting hunting pushed the Tasmanian tiger to the brink of extinction by the turn of the twentieth century, and eliminated it from the wild by the 1920s. The last known thylacine died at Hobart Zoo in 1936, though alleged sightings persist to this day.[23]

What has been the wider impact of species loss and in what ways has this affected human beings? While older environmental histories tended to focus on the consequences of biodiversity loss at a species level, more recent works have gone further in highlighting the impacts of depopulation/extinction for entire ecosystems. If animals that play an important role in plant pollination or seed dispersal decrease or disappear, for example, vegetation may change or die out. If a key prey species diminishes, this will have repercussions all the way up the food chain. If animals that contribute to the shaping of the landscape (such as beavers) are removed, river courses may change and flooding may ensue. And if predators are culled, their prey may multiply unchecked, causing overgrazing and soil erosion. Writing in 1892, for instance, Mr A.C. Macdonald from the Cape Colony Agricultural Department complained that the extermination of aardvarks in the region for their meat and hides had resulted in an explosion of the white ant, 'an insect which feeds on our crops and the succulent herbage of the veld, and which does much greater damage than is generally supposed'.[24] The loss of a single keystone species could thus have a ripple effect across the local environment, impacting the balance of nature and, in some cases, affecting human livelihoods.

To illustrate the wider significance of species loss, let's take two further examples: the sea otter and the vulture. The sea otter is a marine mammal, native to the coastal waters of the north Pacific. Numbering between 150,000 and 300,000 in 1700, sea otter populations were decimated in the mid-eighteenth century by Russian hunters, who sold their thick fur coats to buyers in China. This was clearly bad for sea otters, who could not withstand this unprecedented onslaught on their numbers. It was good, however, for crustaceans and sea urchins, which multiplied in the absence of their only major predator. The increase in crustaceans was, in turn, bad for kelp, which the former consume, and, ultimately, for fish, which, with less kelp to protect them from predators, quickly declined in number. The decline in fish populations subsequently affected the native people of the Aleutian Islands, who relied on fish for sustenance, making the latter more vulnerable to diseases introduced to the region by Europeans. Sea otters can therefore be seen as a keystone species in the North Pacific Ocean, whose removal from the food chain left the wider ecosystem impoverished in terms of species variety and overall biomass – a process ecologists refer to as a trophic cascade.[25]

The vulture offers a particularly stark illustration of how the decline of a single family of animals can impact upon an entire ecosystem – including humans. Once plentiful on the Indian subcontinent, vultures have played a key ecological role by disposing of dead cattle – livestock not eaten by vegetarian Hindus, but kept widely throughout India as working animals. Since the 1970s, however, these carrion-eating birds have been almost completely exterminated by the increased use of the antibiotic diclofenac, which is used to treat lameness and mastitis in cattle, but causes kidney failure when ingested in large quantities. Poisoned by the carcases they ate, populations of the three main vulture species in India fell by 97% between 1992 and 2007 – a catastrophic collapse. As a result, dead cattle have been left to rot, giving rise to bacterial diseases like anthrax and polluting fields and waterways. Other scavengers, meanwhile – especially feral dogs – have increased in number, exacerbating the spread of rabies in canines and humans. Human bone collectors, conversely, who previously eked out a living by gathering de-fleshed cattle bones for fertiliser, have lost their jobs, sinking further into poverty. The destruction of one crucial bird family has thus had major ecological and economic repercussions for both animals and humans – a compelling illustration of species interdependence.[26]

Conservation

So far, this chapter has painted a bleak picture of the impact of humans on other species – and rightly so. But that is not the whole story. Though directly or indirectly responsible for many animal extinctions, humans have also sought to preserve some threatened species, taking measures to prevent their disappearance. Their efforts to do so reflect shifting attitudes towards nature from the late eighteenth century onwards and a growing understanding of humanity's impacts upon the natural world. They also, however, reflect the contradictions and limitations of human compassion for animals, and the degree to which conservation – like natural history – has been inflected by prevailing class and race prejudices.

Humans have long recognised the link between human activities and the decline in animal populations. Before 1800, however, most people were unable to countenance extinction, since the idea that one of God's perfectly created species might completely disappear from the earth's surface was anathema to prevailing biblical understandings of nature. According to Christian teachings, nature was not something unstable and changeable, but a static, orderly, balanced entity in which every species formed a vital and unbreakable link within a wider 'Great Chain of Being'. When evidence was presented of a species's decline or disappearance, therefore, naturalists concluded that the animal in question had simply retreated to more distant parts of the globe where humans had yet to penetrate. Confronted with the bones of a mastodon in 1780, the American politician and naturalist Thomas Jefferson insisted that there must still be living descendants of the beast somewhere in the unexplored interior of the continent, for 'Such is the oeconomy [*sic*] of nature, that no instance can be produced of her having permitted any one race of her animals to become extinct; of her having formed any link in her great work so weak as to be broken'.[27]

Around 1800, this perception of the natural world started to change as evidence of extinction became overwhelming. The discovery of the bones of giant ground sloths, mammoths, mastodons, and other creatures for which no living counterpart could be found forced naturalists to accept (often reluctantly) the possibility that species could die out.[28] From the 1830s, meanwhile, a series of studies on island-dwelling birds such as the moa, the great auk and the dodo proved that these creatures no longer existed anywhere on the globe and, more significantly, that they had likely been exterminated by humans – in the case of the moa, by Maori settlers in New Zealand; in the case of the great auk and the dodo, by European sailors, who killed the birds for their fat, feathers and eggs.[29] With ever more animals pushed to the brink by the end of the nineteenth century – including once numerous species like the bison, the quagga and the passenger pigeon – the human causes of biodiversity loss became even more palpable, prompting concerns that some species would soon follow the dodo into oblivion if nothing were done to curb their exploitation. Writing in 1913 – one year before the last passenger pigeon died in Cincinnati Zoo – a contributor to *The Animal World* listed several animals believed to be 'verging on extinction', among them the Siberian sable and the Andean chinchilla, slaughtered for their fur, the thylacine, killed by farmers for its alleged depredations upon sheep, the flightless kakapo from New Zealand, under threat from introduced predators, and the 'inoffensive' koala, 'in much demand as a pet, owing to its chubby appearance and amiable disposition'.[30] From a conceptual impossibility, therefore, extinction had metamorphosed into an alarming reality – one that required swift action if it were to be delayed or reversed.

What measures, then, have humans taken to ensure the survival of other animal species and how have the strategies and priorities of conservation changed over time?

In the nineteenth century, when conservation first began to be discussed, the emphasis was primarily on over-harvesting – the most obvious and immediate cause of animal depletion. Early proponents of animal preservation duly advocated three primary tactics: hunting bans (or, more commonly, the issuing of hunting licences, to regulate the slaughter), the creation of special reserves in which certain species could not be hunted and the imposition of export and/or import bans on products derived from threatened species. In 1896, for example, German colonial authorities in Dar-es-Salaam issued a game ordinance prohibiting the hunting of elephants without a licence (for which a fee had to be paid).[31] In 1899, ex-big game hunter Alfred Sharpe proposed banning 'the export of [elephant] tusks under a certain weight – say 14lb [6.4kg]' to prevent the killing of female and juvenile elephants for their ivory.[32] Aimed primarily at preserving valued species for continued exploitation – whether as sources of animal products or quarries for big game hunters – most of these measures were utilitarian in design, calling not for a complete cessation of hunting, but for its stricter regulation. As the century drew to a close, however, this highly anthropocentric approach to conservation was challenged by a new emphasis on human stewardship over the natural world, which increasingly advocated the protection of other species for their own sake (or, more specifically, because humans valued their existence). Calling for the preservation of the Indian rhinoceros, for example, American conservationist William Temple Hornaday characterised the animal as 'a gift handed down to us straight out of the Pleistocene age, a million years back' – a beautiful thing to be prized in its own right, and not for what it could do for man.[33]

Twentieth- and twenty-first-century conservation efforts have built on their nineteenth-century predecessors, expanding the range of protected species and adding new tools to the conservation arsenal. Alfred Sharpe's plan to curtail ivory exports, for example, has been extended to multiple other species through the Convention on International Trade in Endangered Species of Wild Fauna and Flora (CITES) (1975), which prohibits trade in animal populations classed as vulnerable. Anti-poaching schemes have been enhanced by new technologies such as camera traps, radio collars and GPS tracking devices, while more sophisticated understandings of the habits and breeding patterns of wild animals have allowed for conservation strategies better tailored to promote their survival. Since 2018, for instance, the conservation charity Pantera has been working to create a Jaguar Corridor across Central and South America, connecting fragmented jaguar populations from northern Mexico to Argentina and preserving the species's genetic integrity.[34] Captive breeding programmes, international stud books for endangered species and novel techniques such as artificial insemination have also played a part in the conservation landscape, contributing to ambitious reintroduction and re-wilding projects. In 1992 scientists working for the US Fish and Wildlife Service released two California condors, Chocoyens and Xewe, into the wild after successfully breeding 25 of the almost extinct species in San Diego Zoo.[35] In 1997 conservationists from the Duke University Primate Center released five ruffed lemurs into the Betampona Reserve in Madagascar, placing the animals in a special 'habituation cage' and gradually reducing their rations of 'bananas, mangoes, pineapple and monkey chow' to encourage them to forage on native forest fruits.[36] While nineteenth-century conservation typically focused on preserving a single species, twentieth- and twenty-first-century conservation has focused increasingly on

preserving entire ecosystems, showing greater sensitivity to the complex relationships between predators, prey, plants, parasites and pollinators.

Since 1800, therefore, humans have shifted from denying the possibility of
extinction, to trying to prevent it. How successful have they been in doing
so? What factors have limited the effectiveness of conservation measures and
what does the shifting nature of conservation priorities reveal about human
attitudes towards other animals – and, indeed, other people?

If we focus, first, on the practical challenges of protecting animals, these
have been extensive, with the result that many conservation efforts have
failed. In order to achieve their intended outcome, conservation measures
such as sanctuaries and export bans required proper policing and monitoring, which in turn required proper funding. Both were often lacking, allowing poaching to continue. As the early nineteenth-century conservationist
S.H. Whitbread remarked in 1907:

> It is not enough to colour spaces on a map and to add in a footnote
> 'spaces coloured pink are Game Reserves'. Reserves must be watched
> and policed to ensure that their limits are maintained inviolate and
> their regulations observed; this means men, and men means money.[37]

For animals such as birds and marine mammals that migrated over long
distances, furthermore, international collaboration was essential to ensure
that vulnerable species enjoyed protection in all the parts of their habitat.
Seals, whales and dolphins, for instance, traversed the waters of multiple
nations. Many bird species migrated between the northern and southern
hemisphere on a seasonal basis, while even terrestrial species such as elephants and giraffes often straddled artificial human frontiers.[38] To ensure
that these animals were adequately protected, transnational efforts were
needed to curb hunting, deforestation and pollution, as well as to combat
the trade in animal products. As Raf de Bont has shown, such international
collaboration started to become more common in the 1920s and 1930s, with
the establishment of increasing numbers of NGOs focused on the preservation of nature.[39]

Alongside practical challenges, animal conservation schemes also pose
complex ethical issues, sometimes compromising the wellbeing of individual
animals for the supposed greater good of the species (a topic we'll explore
further in Chapter 7). In order to relieve pressure on threatened native species, for example, scientists have advocated the culling of non-native 'pests',
from camels and cane toads in Australia to goats on the Galapagos. These
animals are also sentient beings, however, arguably with rights of their own,
so such policies have proven controversial (especially when their targets are
'cute' animals like cats or squirrels). Less overtly violent, but equally notable,
have been the liberties taken with captive-bred animals as part of planned

reintroduction programmes. Some of these animals have undergone artificial insemination or other forms of intensive breeding, while many have perished shortly after release, due to their lack of preparedness for life in the wild. This has undoubtedly caused suffering to the individuals, even if the ultimate conservation goals have been attained. Of fifteen golden lion tamarins released into the Atlantic Forest in Brazil in 1984, for instance, only two survived, some being returned to zoos, others dying after consuming poisonous fruit and yet others dying from exposure because they did not know how to nest in tree holes at night.[40] Saving a species could thus involve significant sacrifices for individual animals, whose suffering was adjudged necessary for the wider good of the programme.

Three further points should be made regarding the overall ethos of conservation, one in relation to its significance for animals, one in relation to its impact on humans and the third about the wider scope and purpose of conservation in the modern world. First, as regards animals, it should be emphasised that conservation has often been highly selective, and that decisions over which species should be 'saved' have primarily reflected human priorities. When the first game regulations were introduced into North-eastern Rhodesia in 1900, for example, they explicitly protected vultures, secretary birds and owls, 'on account of their usefulness', and giraffes, gorillas, mountain zebras and 'the little Liberian hippopotamus', 'on account of their rarity and threatened extinction', but they excluded lions, leopards, poisonous snakes, crocodiles, baboons 'and other harmful monkeys' – none of which was useful to human beings.[41] In 1924, meanwhile, when British conservation advocate Albert Gray called for tighter protections for the gorilla and other great apes – whose killing he equated to 'murder' – he received a mixed response from contemporaries, some of whom were lukewarm about Africa's primates.[42] One commentator, David Wilson-Barker, insisted that 'there are some places where monkeys (especially baboons) are a positive nuisance and must be kept down, or it would not be possible to raise crops'. Another, Sir Arnold Hodson, recommended lesser protections for chimpanzees, which, in Sierra Leone, were alleged to be 'baby killers, taking the children up to the tops of trees and throwing them to the ground'.[43] While such overtly anthropocentric judgements on other species have become less common since the mid-twentieth-century, human biases have nonetheless continued to skew conservation priorities, with so-called 'charismatic megafauna' attracting the most interest – and, crucially, the most funding. Pandas, for instance, have elicited a disproportionate amount of support in China (and beyond), due to their (comparatively recent) elevation to the role of Chinese national animal, while giant tortoises have become icons of conservation in the Galapagos, owing to their links to Charles Darwin and the study of evolution.[44] Amphibians, fish and insects, by contrast, have struggled to generate similar levels of public concern, and, in consequence, have received less funding for their preservation; Hawaii's native snail species, though critically endangered, have attracted substantially less federal funding than vertebrate species in an equivalent position.[45] When it comes to conservation therefore, animals have tended to be valued according to human standards, not on their own merits (though, of course, the appearance and behaviours of different species have influenced these perceptions).

Second, it is important to note that conservation measures have often been highly discriminatory towards certain groups of humans, particularly when enacted in colonial or postcolonial contexts. In late nineteenth- and early twentieth-century North America, the establishment of national parks at Yellowstone (1872) and the Grand Canyon (1893) excluded Native American peoples like the Crow, the Shoshone and the Havasupai from their traditional hunting grounds and criminalised the seasonal killing of deer and bison.[46] In colonial Africa, meanwhile, imperial conservation legislation was explicitly targeted at the indigenous African population, who were accused of using cruel methods (such as pitfalls) to kill large game, prevented from purchasing hunting licences by high fees and, later, forcibly displaced from their homes to make way for game reserves; in the 1920s and 1930s thousands of families from the Banyarwanda, Nande and Heme populations were evicted from the newly created Albert National Park in the Belgian Congo, while other groups living outside the park were forbidden to hunt, collect bamboo or graze cattle within its boundaries.[47] Conservation has thus often gone hand-in-hand with imperialism and class discrimination, reinforcing existing social hierarchies and perpetuating inequalities. Though less blatant in the twenty-first century, such attitudes still prevail in some quarters, leading to accusations that (often white, middle-class) conservationists prioritise the survival of wild animals over the preservation of human life. Balancing social justice and nature protection thus remains a challenge in the modern world, as growing human populations compete with other species for dwindling natural resources.

Finally, in assessing the effectiveness of conservation initiatives, it is important to think more deeply about the objectives behind such schemes and to ask what 'success' actually looks like. Is the purpose of conservation to maintain animal populations at their current level, preventing further losses, or is it to restore nature to some earlier, 'pristine' state before human actions began to destabilise it? If the latter, when should that time be, and should conservation entail bringing extinct species back into being (a process that has been attempted on several occasions, with varying degrees of success), or substituting extinct species with other, similar animals that are capable of performing the same functions within a given ecosystem? Are such ends possible, or even desirable? Historians have contributed meaningfully to these debates by bringing a historical perspective to scientific discussions. Charting conservation efforts on the Galapagos, Elizabeth Hennessey shows how twentieth-century conservationists attempted to return the fauna and flora on the islands to the state they were in in 1534, the year before they were first 'discovered' by a Spanish bishop, Fray Tomás de Berlanga. She argues, however, that this 'return to Eden' was unattainable, owing to the practical impossibility of eradicating all invasive species, the change in distribution of native species that had occurred in the intervening centuries and the extinction of some species of tortoises, who could no longer return to recolonise their former homelands.[48] Looking at attempts to reanimate the extinct African quagga, Sandra Swart questions both the legitimacy and the ethics of 'de-extinction' efforts and asks whether 'Kevin', a 'quagga' born in 2009 from three decades of zebra crosses, is a genuine resurrection of a species not seen alive since 1883, or merely a zebra with fainter-than-average stripes (he may look like a quagga, but does he behave, move or sound like one?). Even if we accept Kevin as an authentic quagga, Swart questions the morality of bringing extinct species back from the dead. Are we doing so for their benefit or ours, to restore ecosystems or merely to satisfy scientific curiosity and provide a spectacle for tourists? Is there any point in resurrecting extinct animal species if the conditions that brought about their extinction have not changed (a question that also applies to reintroduction and rewilding efforts)?[49] Conservation, therefore poses a range of challenging logistical and ethical questions and needs to be seen within its own social and historical context.

Ecological Indians?

The previous sections have focused primarily on changing European attitudes towards wildlife. This, however, raises an important question which environmental historians are starting to probe more deeply: Were Europeans the sole perpetrators of biodiversity loss, or have other non-European humans also contributed to the destruction of animal species? Was there something specific about European culture and values that encouraged the unrestrained exploitation of other living things? How accurate is the stereotype of the so-called 'Ecological Indian' – the non-Western (usually Native American) proto-conservationist who lived in harmony with nature and managed natural resources with restraint?

A survey of man's relationship with animals across the centuries suggests that Europeans and their descendants have inflicted the greatest damage on animal life. Two key elements of European society appear to have been primarily responsible for this destruction: Christianity and capitalism. The former emphasised man's dominion over animals as one of its central teachings, propagating the view that animals were made for man's use and could (indeed should) be exploited for his benefit (though how far this shaped actual human–animal interactions has been debated). The latter fostered the idea of endless wealth accumulation, encouraging humans to exploit animals (and other natural resources) beyond mere subsistence level. Together with new technologies – notably transoceanic shipping and firearms – these acquisitive tendencies have pushed a range of animals to the brink of extinction, first in Europe and, later, in colonised territories. Steller's sea cow was exterminated within just three decades of its 'discovery' by Russian explorers, while the bison and the egret were both, in different ways, victims of European-centred commodity booms, the former for robes, hides and fertiliser, the latter for feathers, used as adornment for women's hats. Japan's wolf population went extinct following the Meiji Restoration of 1868 and the subsequent adoption of Western livestock-raising practices. In all these cases, contact with Europeans seems to have been a crucial tipping point after centuries of apparent coexistence with non-European peoples.

By contrast, it has been suggested that some non-European societies – especially the indigenous peoples of the Americas – had a less exploitative relationship with nature, the result in part of belief systems that drew a less sharp distinction between humans and the natural world. Where Christianity placed humans on a separate level from other animals and permitted their exploitation, many Native American religions stressed the importance of balance in one's use of nature, and the need to avoid excessive extraction for fear of spiritual repercussions. This made them less inclined to kill more animals than they needed for sustenance and more inclined to respect other species, which some saw as their close relatives. As Shawn William Miller expresses it: 'Nature, for Indians, was a power to be reckoned with, equal to or greater than human powers'.[50] Where capitalism encouraged unlimited consumption, moreover, and the constant search for new markets, some hierarchical indigenous societies such as the Aztecs and the Incas introduced strict sumptuary laws, restricting the wearing of furs and feathers to a small elite and thereby reducing demand for such products (see Chapter 1). The result of these beliefs and practices was that Native American hunting of animals was, in at least some cases, more sustainable than its European counterpart, leaving a lighter ecological footprint. John Richards, for example, suggests that indigenous whale hunting in the Arctic in the period 1200–1600 probably accounted for no more than 120 bowhead whales per year, making only a 'slight impact on the stocks of whales' in the region.[51]

While the stereotype of the 'Ecological Indian' thus has some foundation in fact, recent historical studies have demonstrated its limitations, undermining the idea that all indigenous engagement with nature has been benign. First, as historians have pointed out, species extinctions have sometimes predated the arrival of Europeans, showing that other civilisations were also capable of ecological destruction. The ancient Hawaiians (*c*.400–1810), for example, cleared lands for agriculture, diverted streams and introduced dogs, pigs, chickens and rats to the island, resulting in the extinction of at least forty indigenous bird species. Polynesian hunters eradicated thirteen different species of moa in New Zealand before any European set foot in the country. In Holocene South America, meanwhile, overhunting appears to have played a part in the extinction of megafauna like the glyptodon (a species of giant armadillo) and the megatherium (a type of giant ground sloth), both of which disappeared from the continent at some point between 11,000 and 10,000 years ago.[52] While Europeans undoubtedly extinguished species at a faster rate, therefore, many animals had already come under pressure from non-Europeans by the time the first white settlers reached their homelands; some had disappeared completely without European intervention.

Second, though some non-European peoples doubtless engaged in a more measured relationship with the natural world than their European counterparts, there are clear examples of killing to excess, either for sacrifice or personal adornment. Aztec tribute lists include vast quantities of animals, living and dead, from ocelots and jaguars for the emperor's menagerie to deer, rabbits and quetzal birds for their flesh and feathers. In Peru, meanwhile, the conquistador Pedro Sancho counted 100,000 dead birds in one Inca storeroom.[53] This suggests that at least some of America's indigenous people consumed animals in excessive numbers, even if such consumption was confined to a privileged elite. Nor, moreover, were indigenous peoples immune to wasting natural resources, for some clearly killed more animals than they needed for their subsistence. Writing in 1857, Henry Hind claimed to have witnessed a Cree pound strewn with the remains of bison, each of which had been 'deprived of its tongue and hump only', with the remainder of the carcase left to rot.[54] Forty-two years later, Alfred Sharpe, governor of Nyasaland (modern-day Malawi), contended that 'the African native throughout the continent since the introduction of firearms, urged on by the high value of ivory in European markets, has slaughtered elephants wherever he could find them, regardless of size or of sex; and so long as ivory of all descriptions is a valuable trade article, elephants will continue to be indiscriminately killed, until, in many portions of Africa, they will be totally exterminated'.[55] European markets were clearly in part culpable for this destruction, but Sharpe (admittedly not an impartial witness), viewed the native peoples as complicit in the slaughter.

Lastly, it should be noted that, while the killing of animals among non-European societies was often framed by rituals and cultural taboos, this did not necessarily limit consumption of other species and could, in some cases, encourage it. The North American Cherokee, for example, believe that they must ask for an animal's pardon before killing it, otherwise they will be afflicted by rheumatism; the Cree must place beaver bones in water to allow them to reincarnate. Though a sign of respect for the hunted animal, such a belief in reincarnation, or the idea (common among Native Americans) that animals are voluntarily surrendering themselves to hunters, might increase the level of such killing rather than acting as a prompt for conservation, since it could foster the view that animal populations are limitless. As Shepherd Krech III notes:

> The Yupit of south western Alaska ... thought that the more meat they consumed and shared, the more they would have; that animals would regenerate infinitely as long as they received proper respect from men; and that animal populations declined from lack of respect, not overhunting.[56]

Conversely, in other situations, negative beliefs about particular species have resulted in their destruction for purely spiritual reasons, often with devastating consequences. In twentieth-century Madagascar, for instance, a long-standing belief in the malevolence of the aye-aye (a nocturnal species of lemur) has caused many Malagasy to kill the animal on sight, severely depleting a species already under pressure from deforestation.[57] Not all non-European interactions with animals were therefore benign, and even positive relations often stemmed more from fear than from respect and friendship.

Overall, then, the picture turns out to be more complicated than it might first have appeared. Some non-European peoples have acted as stewards of nature, while others have not. Yet others have embraced new technologies and hunting practices following exposure to European colonialism.[58] While Europeans have clearly inflicted immense damage on the natural world, therefore, it is important not to generalise across other non-European groups and not to present all native peoples as a homogenised whole. It is also important to recognise that tribal peoples have agency and can evolve – their cultures are not frozen in time, and their perceptions of nature may change in response to social shifts or wider market forces. This can have significant implications for their relationship with other animals.

Animals Imperilled – and Saved (?)

To explore the complex ways in which human actions have contributed to the decline and survival of non-human animals the chapter concludes with case studies of two very different species: the Peruvian vicuña and the New Zealand kakapo. Perfectly adapted to their native environments, both species suffered major declines as a direct and indirect result of human contact. Both, however, have been brought back from the brink by subsequent human intervention. Their experiences provide useful insights into the diverse causes of biodiversity loss, the respective impacts of European and non-European actors and the complex ethics of animal conservation.

The Vicuña

The vicuña is a species of camelid native to Peru and Bolivia. Found at altitudes of 3,700 metres and above, it is supremely well adapted to life in the high *puna*, its oval-shaped blood cells maximising oxygen absorption into the blood stream, and its large heart (almost 50% bigger than the average heart weight for mammals of a similar size), pumping oxygen around its body. It is most famous for its soft, thick coat, which insulates its owner from the cold and protects it from the high UV radiation experienced at high altitudes.

Vicuñas were hunted for their wool prior to the Spanish conquest of Peru and especially valued by the Incas, who used vicuña fibres for their most exquisite textiles. According to Jesuit missionary Bernabe Cobo, the clothes of the Inca Emperor 'were made entirely or partially of vicuña wool', mixed with 'viscacha wool, which is very thin and soft, and bat fur ... which is the most delicate of all'.[59] The Incas did not farm vicuñas, as they did alpacas, but hunted them intermittently in drives known as *chakkus* – the sustainability of which was open to question. Spanish conquistador Pedro Cieza de León estimated that 'as many as thirty thousand head' of vicuñas might be killed during a single *chakku* – a very substantial number.[60] Mestizo chronicler Garcilaso de la Vega, however, claimed that only the males perished in these hunts, for the Incas would release female vicuñas 'after they had been sheared'. He also stated that the Incas kept records of the number of animals caught and killed, and would only hunt in a particular region one year in every four.[61] The Inca *chakku* thus appears to have been a sustainable activity 'regulated by political, religious, and other local social and cultural mechanisms'.[62]

This state of affairs changed following the arrival of the Spanish in 1532. Where the Incas had hunted the graceful camelids intermittently, in carefully controlled drives, the Spanish pursued them without remission, shooting them with guns and running them down with dogs and horses (Figure 6.1). Spanish botanist Hipólito Ruíz, who witnessed a vicuña hunt in Peru in the 1780s, described how local people trapped vicuñas in an enclosure, 'shouting, beating drums, blowing whistles, and snapping whips' to intimidate the animals and running them down on horseback.[63] Half a century later, the Swiss naturalist Johann Jakob von Tschudi recorded details of a vicuña hunt he observed in the Altos of Huyhuay, in which 122 vicuñas were herded into a corral and killed.[64] While the Spanish initially exploited vicuñas for the bezoar stones often found in their intestines, by the eighteenth century attention had shifted to the camelid's exquisite fleece, used to make scarves, gloves, stockings, hats, sheets and handkerchiefs. Thousands of vicuña skins were sent to Spain to meet the growing demand, and thousands of vicuñas perished in the process. One contemporary calculated that around 100 arrobas (approx. 1,134 kilos) of vicuña wool were imported into Spain every

Figure 6.1 'The Mêlée: Scene in Aparoma', April 1857. *Series 01: Annotated Watercolour Sketches by Santiago Savage, 1857–1858, being a Record of Charles Ledger's Journeys in Peru and Chile; with maps and notes.*

Source: Mitchell Library, State Library of New South Wales MLMSS 630/1.

year by the turn of the nineteenth century, which, at 12 ounces of wool per vicuña, equated to 8,000 dead vicuñas.[65]

With such large numbers of vicuñas now being killed, it became apparent to the Spanish Crown that reform was needed to ensure the long-term viability of the trade in their wool. This gave rise to two key conservation policies: domestication and hunting legislation. Domestication, reflecting prevailing notions of agricultural improvement, centred on the idea that the vicuña, a wild animal, might be effectively tamed and farmed for its wool in a similar way to its close relative, the alpaca, thus rendering its exploitation sustainable. It found favour with several nineteenth-century reformers, who hoped both to preserve the vicuña and to maximise its profitability. In 1810, for instance, the botanist Francisco José de Caldas advocated shipping one thousand vicuñas to the mountains of the Sierra Nevada in his native New Granada (modern-day Colombia), where they could be intermittently rounded up for shearing.[66] In the 1830s and 1840s a Peruvian citizen, the parish priest Juan Pablo Cabrera, started an experiment to breed vicuñas with alpacas in the remote village of Macusani, keeping the animals in a corral, to prevent them from breeding with wild male vicuñas, and feeding them on 'bread, cake, coca leaves, sugar [and] corn on the cob'.[67] Sadly, none of these schemes came to fruition, leaving the vicuña vulnerable to continued overharvesting.

Hunting legislation for the protection of the vicuña first appeared on the statute books in the sixteenth century and re-surfaced intermittently throughout the colonial period. The first formal conservation measure came in 1557, when the Spanish Crown imposed a five-year moratorium on vicuña hunts to allow the population to recover. Further hunting bans followed in 1768, 1777 and 1789, under the new Bourbon dynasty, while the Liberator Simón Bolívar reiterated the prohibition in 1825, shortly after securing Peru's independence from Spain.[68] Twenty years later, in 1845, the Peruvian Government replaced Bolivar's order with a more comprehensive frame-work of legal protection, designed not only to prevent the over-hunting of vicuñas, but to maintain the nation's monopoly over alpacas. The new law stipulated that 'the exportation of live vicuñas and alpacas to foreign countries' was 'absolutely prohibited'; that earlier Spanish decrees banning the killing of vicuñas would be reinstated; and that the authorities would prosecute anyone found continuing the 'barbarous custom' of killing vicuñas in 'traps' or 'with dogs'.[69] In 1907 Peru banned the export of vicuña skins completely. In 1918 neighbouring Bolivia followed suit.[70]

Despite these measures, the vicuña was critically endangered by the mid-twentieth century, with fewer than 6,000 left in the wild by 1960. This prompted a final and ultimately highly effective effort to save the species which saw vicuñas placed in Appendix I of The Convention on International Trade in Endangered Species (CITES) and an export ban imposed on their fleece. Aware of the need for international cooperation in conservation, the governments of Peru, Bolivia, Chile and Argentina issued legislation to pro-tect vicuñas within their territories and signed an international agreement in 1979 to co-ordinate conservation efforts. In 1967 the Peruvian Government established the first of several special reserves for vicuñas at Pampa Galeras, Ayacucho province, where the animals could be monitored by scientists and protected from poachers. Thanks to these conservation measures, the decline in the vicuña population was reversed and their numbers rapidly increased – even leading to debates at the end of the 1970s over whether some of the animals should be culled or translocated. The guerrilla war unleashed by *Sendero Luminoso* in the 1980s constituted a temporary setback for vicuña conservation, with poaching once again unchecked and the reserves unpo-liced, but by the early 1990s the species was in recovery, and vicuñas were downgraded to Appendix II on CITES in 1994.[71] Classed as a Peruvian heri-tage species, the vicuña still cannot be hunted, but it can be shorn and released by indigenous people using the traditional Inca *chakku* technique.

The fluctuating fortunes of the vicuña highlight the degree to which over-harvesting could put pressure on wild animals, but also the capability of fast-reproducing species to bounce back when given protection. Managed in a seemingly sustainable manner by the Incas and other pre-Columbian peo-ples (in part due to the existence of strict sumptuary rules), vicuñas were slaughtered without restraint following the arrival of Europeans, suffering a catastrophic drop in numbers. Repeated conservation efforts initially met with limited success, primarily due to ineffective enforcement. From the mid-twentieth century, however, their effectiveness began to increase, and the vicuña population has since rebounded, reaching roughly 340,000 by 2015. While poaching, habitat loss and climate change continue to pose problems for the vicuña's long-term survival, its rehabilitation is widely regarded as a conservation success story, and a model that might be applied to other species. South Africans advocating the farming of rhinos for their horns cite the management of the vicuña as an example of successful conservation through sustainable use, though vicuña experts are cautious about the extrapolation of this practice to another, very different, species.[72]

The Kakapo

The kakapo is the world's heaviest parrot, and the only parrot species that cannot fly. It is nocturnal (kakapo means 'night parrot' in Maori), can live for up to sixty years and breeds only every two or three summers, depending on the amount of nutritious rimu berries it can find. Male kakapos are significantly larger than females and attract mates by constructing bowls in the soil and emitting a distinctive booming call. Both sexes sport a mottled green plumage which camouflages them effectively from native birds of prey (Figure 6.2).

Kakapos are indigenous to New Zealand, which, due to its geographical separation from other continents, lacks any native mammals (apart from bats). For millennia, the parrots thrived in this environment, maximising their ecological niche. As soon as humans arrived on the islands, however, that situation began to change. The arrival of Maori settlers from Polynesia between 1320 and 1350 led to a significant decline in the kakapo population, which was eliminated from many parts of the islands by hunting with dogs (for their skin and feathers), deforestation and depredations on their eggs and chicks by the Polynesian rat (the kiore). Five hundred years later, the arrival of Europeans in the 1840s accelerated this decline, increasing levels of hunting and deforestation and exposing the parrot to a new influx of foreign predators, in this case cats, black rats, stoats and ferrets (the latter introduced in the 1880s to reduce the rabbit population).[73] Less directly, the kakapo suffered from the introduction of European deer, which competed with the parrot for food. As its numbers dwindled, it also fell victim to scientific collectors, who wanted to secure the last remaining specimens for museums. An article from 1913 commented that: 'Although formerly the kakapo was to be found over the whole of New Zealand, yet its range is now so restricted, and its numbers so diminished that it is in danger of becoming extinct at a very early date'.[74]

Figure 6.2 'Papageien', from *Meyers Konversation-Lexicon*, 4th Edition (Leipzig: Bibliographisches Institut Leipzig, 1898), Plate 1.

Source: photograph by author.

Efforts to conserve the kakapo began in the late-nineteenth century, when its decline was first noted, and concentrated on relocating survivors to island sanctuaries. In 1897, for instance, naturalist Richard Henry transferred 135 kakapos from the mainland to Resolution Island and established a protected colony there.[75] Unfortunately, stoats swam out to the island in 1900, killing most of the kakapos by 1906. A second, smaller kakapo colony was established on Little Barrier Island but suffered the same fate, this time at the hands of feral cats. A third effort to save the kakapo in the mid-twentieth century also ended in failure, with four out of five kakapos relocated to an isolated island in 1961 dying within months of capture. By 1996, there were only 47 kakapos alive in the wild.[76]

From this low point, the kakapo's fortunes began to turn, thanks in large part to the advent of a special Kakapo Recovery Plan by the New Zealand Government. Building on previous conservation strategies, scientists once again collected the surviving kakapos and relocated them to several uninhabited islands. Conservationists used new technologies such as radio transmitters to monitor the birds and made persistent efforts to eradicate potential predators such as rats, feral cats and stoats. Scientists also went to quite extreme lengths to foster kakapo reproduction, extracting semen from male birds to artificially inseminate the females; supplementing the diets of females with apples, sweet potatoes, almonds, Brazil nuts, sunflower seeds and walnuts to increase fertility; developing realistic fake eggs fitted with smart technology so that the real ones could be taken out of the nest for safer incubation, (because female kakapos sometimes accidently crush their own eggs); and even, in one case, repairing a crushed egg with sticky tape to facilitate the hatching of a chick.[77] A hand-reared male parrot named Sirocco, meanwhile, became a global ambassador for the species in 2009 after footage emerged of him attempting to mate with the head of a zoologist (he had been raised by humans and imprinted on them), generating widespread interest in (and funding for) the conservation of his fellow parrots. By 2023, thanks to these efforts, kakapo numbers had risen to 247 – still tiny, but promising.[78]

The kakapo's story is one of devastating decline and recent (tentative) conservation success. Perfectly adapted for the conditions of its isolated homeland, this nocturnal, land-dwelling parrot was an early victim of external colonisation, first by the Maori, later by Europeans. Hunting for food and adornment made an immediate impact on kakapo numbers. Deforestation rendered the species more vulnerable to attack, while invasive species did the most damage, killing adult kakapos and destroying their eggs. More recently, a deadly fungus has attacked surviving kakapos, underlining, once again, the dangers of species interchange in an increasingly globalised world.[79] The story of the kakapo thus highlights the multicausal nature of species decline, the importance of global migration in extinguishing vulnerable creatures – especially in isolated island ecosystems – and the fact that species extinction has not been solely caused by Europeans (though in this case they did significantly accelerate it). It also illustrates the challenges, and eventual (partial) effectiveness of species conservation, which, for this unusual parrot, has been taken to extreme levels.

The fate of the kakapo also raises interesting questions relating to the wider ethics of conservation. First, is it right to subject individual kakapos to artificial insemination and other invasive measures in order to save their species? In the case of Sirocco, for instance, scientists attempted to manually collect semen from the bird (which he resisted), created a motorised decoy female kakapo to coax him into mating (which failed) and even developed a special 'ejaculation helmet' for humans to wear, so that semen could be collected when the parrot attempted to mate with human heads (this also failed). Another kakapo, Richard Henry, was subjected to equally invasive procedures in the 1980s. He was known for being 'a very placid bird', but his temper soured in 1987, after 'he was caught so a blood sample could be taken for research comparing Stewart Island and Fjordland kakapos'.[80] Second, does Sirocco's fame trivialise the plight of the kakapo, and is it good to anthropomorphise wild animals in this way? Third, is it justifiable to exterminate other living things (in this case, cats, rats and stoats) in order to ensure the survival of a specific chosen species? Fourth, is it acceptable to concentrate on saving charismatic species like the kakapo at the expense of non-charismatic ones like beetles? Some scientists argue that the two goals are not mutually exclusive, since non-target species will experience a 'trickle-down' effect of preserving key ecosystems. Others, however, assert that channelling funding towards charismatic species is not as effective as supporting other species directly, suggesting that conservation priorities should be reassessed. A recent newspaper article claimed that Sirocco cost his carers a total of $16,633 between 2017 and 2020 – the majority of which was spent on hiring aircraft to transport him around New Zealand for media engagements.[81] Is this the most efficient use of the limited funds for conservation?

Conclusion

Human actions have contributed to biodiversity loss on a global scale. The destruction of animal habitats, the introduction of non-native species, pollution, (human) population growth and the overharvesting of animal products have reduced populations of many non-human animals, pushing some to extinction. Since the late twentieth century, human-induced climate change, micro-plastics and viruses linked to intensive farming have even impacted on animal species with no direct human contact, from Emperor penguins in Antarctica to green turtles in the Pacific Ocean.

While humans have had a predominantly negative impact on the survival of other animals, some humans have made a concerted effort to keep some species in existence, setting limits on hunting, establishing sanctuaries, imposing trade bans on animal products and, more recently, orchestrating animal reintroductions and rewilding initiatives. How successful such strategies have been, and how success should be judged, remain complex issues.

Which species should receive protection (and on what basis)? What does 'saving' a species or 'restoring' an environment truly mean? How can animal conservation be balanced against human needs? In 2024 the president of Botswana, Mokgweetsi Masisi, criticised a proposed British ban on the importation of hunting trophies, arguing that Europeans cared more about elephants than humans – dozens of whom are killed by pachyderms every year in his country. He threatened to send 30,000 elephants to Hyde Park so that the British public could experience what it was really like to live alongside dangerous wild animals.[82] Was Masisi right? Has elephant conservation come at the expense of human safety? Are some animal (and human) lives valued more highly than others?

Notes

1 Ryan Tucker Jones, *Empire of Extinction* (Oxford: Oxford University Press, 2014), pp. 1–20.
2 'Hearing of Manatees May Prove to be Key to Protecting Species', *New York Times*, 25 Augus 1992.
3 'Manatee Rescue Service', *Chicago Tribune*, 22 April 1993.
4 'Manatee Has Spinal Surgery', *Washington Post*, 16 August 1999.
5 The estimated biomass of livestock is twice as much as that of all wild mammals combined. Marcelo Sánchez-Villagra, *The Process of Animal Domestication* (Princeton: Princeton University Press, 2022), p.vii.
6 R.E.A. Almond, M. Grooten, D. Juffe Bignoli and T Petersen (eds), *WWF Living Planet Report 2022* (WWF, Gland, 2022), p. 32.
7 Ursula K. Heise, *Imagining Extinction: The Cultural Meanings of Endangered Species* (Chicago: University of Chicago Press, 2016), p. 22.
8 R.J. Oglesby, T. L. Sever, W. Saturno, D. J. Erickson III, and J. Srikishen, 'Collapse of the Maya: Could Deforestation Have Contributed?', *Journal of Geophysical Research* 115 (2010), D12106.
9 'Trump's Border Wall Construction Threatens Survival of Jaguars in the US', *The Guardian*, 1 December 2020.
10 'Additions to the Zoological Gardens', *The Times*, 3 September 1907.
11 Elinor Melville, *A Plague of Sheep: Environmental Consequences of the Conquest of Mexico* (Cambridge: Cambridge University Press, 1994).
12 David Quammen, *Song of the Dodo: Island Biogeography and the Age of Extinction* (London: Pimlico, 1996), pp. 262–270.
13 Nigel Turvey, *Cane Toads: A Tale of Sugar, Politics and Flawed Science* (Sydney: University of Sydney Press, 2013).
14 'How Endangered a Species?', *New York Times*, 12 February 2002.
15 Brett Walker, *The Lost Wolves of Japan* (Seattle: University of Washington Press, 2005), pp. 129–83.
16 John Soluri, *Creatures of Fashion: Animals, Global Markets and the Transformation of Patagonia* (Chapel Hill: Duke University Press, 2024), pp. 135–141.
17 'Hunting the Panda', *The Times*, 17 January 1939.
18 Hannah Ritchie et al., 'Population Growth' (2023), retrieved from https://ourworldindata.org/world-population-growth.
19 'Ryrkaypiy: Far-North Russian Village Overrun by Polar Bears', BBC News, 5 December 2019.

20 L.M. van Eeden et al., *Impacts of the Unprecedented 2019–2020 Bushfires on Australian Animals* (Ultimo: WWF, 2020), p. 21; R.A. Kock et al., *Science Advances* 4:8 (2018), eaao2314.

21 Andrew Isenberg, *The Destruction of the American Bison* (Cambridge: Cambridge University Press, 2001).

22 Philip Lymbery, *Dead Zone: Where the Wild Things Were* (London: Bloomsbury, 2017), p. 57.

23 Robert Paddle, *The Last Tasmanian Tiger: The History and Extinction of the Thylacine* (Cambridge: Cambridge University Press, 2000).

24 'The Extermination of the Ant-Bear', *The Times*, 24 September 1892.

25 Tucker Jones, *Empire of Extinction*, pp. 97–98.

26 Thom Van Dooren, *Flight Ways: Life and Loss at the Edge of Extinction* (New York: Columbia University Press, 2014), pp. 45–61.

27 Thomas Jefferson, *Notes on the State of Virginia* (Baltimore: W. Pechin, 1800), pp. 54–55.

28 Elizabeth Kolbert, *The Sixth Extinction: An Unnatural History* (London: Bloomsbury, 2014), pp. 23–46.

29 Mark Barrow, *Nature's Ghosts: Confronting Extinction from the Age of Jefferson to the Age of Ecology* (Chicago: University of Chicago Press, 2009), pp. 47–77.

30 'Animals Verging on Extinction', *The Animal World*, February 1913, pp. 27–30.

31 Bernhard Gissibl, *The Nature of German Imperialism: Conservation and the Politics of Wildlife in Colonial East Africa* (New York: Berghahn, 2016), pp. 85–90.

32 Alfred Sharpe, 'The Preservation of African Elephants', *Edinburgh Magazine*, January 1899, p. 91.

33 William Temple Hornaday, *Our Vanishing Wild Life: its Extermination and Preservation* (New York: Clark and Fitts, 1913), p. 189.

34 'Panthera's Jaguar Program', retrieved from https://panthera.org/pantheras-jaguar-program.

35 'Two Condors are Freed; Now Can They Survive?', *Chicago Tribune*, 15 January 1992.

36 'Endangered Mammals Go Home to a Land They Never Knew', *New York Times*, 2 December 1997.

37 S.H. Whitbread, 'The Year', *Journal of the Society for the Preservation of the Wild Fauna of the Empire* Vol. III (1907), p. 12. On species reintroductions and re-wilding initiatives in Scandinavia, see Dolly Jørgensen, *Recovering Lost Species in the Modern Age: Histories of Longing and Belonging* (Cambridge, MA: MIT Press, 2019).

38 Kurk Dorsey, *Whales and Nations: Environmental Diplomacy on the High Seas* (Seattle: University of Washington Press, 2013); Camilo Quintero Toro, *Birds of Empire, Birds of Nation* (Bogotá: Universidad de los Andes, 2012), pp. 131–164.

39 Raf de Bont, *Nature's Diplomats: Science, Internationalism and Preservation, 1920–1960* (Pittsburgh: University of Pittsburgh Press, 2021), pp. 45–84.

40 'Gold Monkeys Learn How to Live in Wild in Brazilian Preserve', *New York Times*, 17 October 1989. See also Monica Vasile, 'From Reintroduction to Rewilding: Autonomy, Agency and the Messy Liberation of the European Bison', *Environment and History* (2022), doi:10.3197/096734022X16552219786618.

41 'North-Eastern Rhodesia', *Journal of the Society for the Preservation of the Wild Fauna of Empire*, vol. II (1905), pp. 74–75.

42 'Hunted Gorillas', *The Times*, 10 June 1924.

43 'Letters to the Editor: Hunted Gorillas', *The Times*, 21 June 1924; 'Protecting African Animals', *The Times*, 7 November 1933.

44 E. Elena Songster, *Panda Nation: The Construction and Conservation of China's Modern Icon* (Oxford: Oxford University Press, 2018), pp. 102–126.

45 Thom van Dooren, *A World in a Shell: Snail Stories for a Time of Extinctions* (Cambridge: MIT Press, 2022), pp. 6–11.

46 Karl Jacoby, *Crimes Against Nature: Squatters, Poachers, Thieves and the Hidden History of American Conservation* (Berkeley: University of California Press, 2001), pp. 81–191.

47 De Bont, *Nature's Diplomats*, pp. 135–167.

48 Elizabeth Hennessy, *On the Backs of Tortoises: Darwin, the Galápagos and the Fate of an Evolutionary Eden* (New Haven: Yale University Press, 2019), pp. 171–179.

49 Sandra Swart, 'Zombie Zoology: History and Reanimating Extinct Animals', in Susan Nance (ed.), *The Historical Animal* (Syracuse: Syracuse University Press, 2015), pp. 54–74.

50 Shawn William Miller, *An Environmental History of Latin America* (Cambridge, Cambridge University Press, 2007), p. 27.

51 John F. Richards, *The World Hunt: An Environmental History of the Commodification of Animals* (University of California Press: Berkeley, 2014), pp. 118–122.

52 Shepherd Krech III, *The Ecological Indian: Myth and History* (New York: Norton, 1999), pp. 29–43.

53 Miller, *An Environmental History of Latin America*, pp. 32–33.

54 Krech III, *The Ecological Indian*, pp. 133–134.

55 Alfred Sharpe, 'The Preservation of African Elephants', *Edinburgh Magazine*, January 1899, p. 89.

56 Krech III, *The Ecological Indian*, pp. 168, 204 and 231.

57 'Lemurs Bred in Captivity May One Day Return to Repopulate Madagascar', *New York Times*, 19 May 1992.

58 Robert Irwin, 'Canada, Aboriginal Sealing, and the North Pacific Fur Seal Convention', *Environmental History* 20 (2015), pp. 57–82.

59 Bernabe Cobo, *Historia del Nuevo Mundo* (Seville, Imprenta de E. Rasco, 1895), vol. IV, p. 205.

60 Clements R. Markham (ed., trans.), *The Travels of Pedro Cieza de León, A.D. 1532–50, Contained in the First Part of his Chronicle of Peru* (London, The Hakluyt Society, 1864), Vol. II, pp. 45–46.

61 Garcilaso de la Vega, *Primera Parte de los Comentarios Reales* (Madrid: Imprenta de Doña Catalina Piñuela, 1829), pp. 447–449.

62 Glynn Custred, 'Hunting Technologies in Andean Cultures', *Journal de la Société des Americanistes*, Vol. LXVI (1979), p. 14.

63 Hipólito Ruíz, *The Journals of Hipólito Ruíz, Spanish Botanist Peru and Chile 1777–1788*, trans. Richard Evans Schultes María José Nemry von Thenen de Jaramillo-Arango (Portland: Timber Press, 1998), p. 104.

64 Johann von Tschudi, *Travels in Peru, During the Years 1838–1842*, trans. Thomasina Ross (London: David Bogue, 1857), pp. 313–314.

65 'De las Vicuñas', *Semanario de Agricultura y Artes Dirigido a los Parrócos*, vol. X, 1801, pp. 263–264.

66 'Memoria sobre la Importancia de Connaturalizar en el Reino la Vicuña del Perú y Chile', in Francisco José de Caldas, *Obras Completas de Francisco José de Caldas*, (Bogotá: Imprenta Nacional, 1966), pp. 323–333.

67 'Noticia exacta del Ingerto de Paco Vicuñas', *El Peruano*, 9 September 1846.

68 Hugo Yacobaccio, 'The Historical Relationship between People and the Vicuña', in Iain Gordon (ed.), *The Vicuña: The Theory and Practice of Community-Based Wildlife Management*, (New York: Springer, 2009), pp. 14–16; Daniel Florencio O'Leary, *Memorias del General O'Leary* (Caracas: Imprenta de El Monitor, 1883), p. 363.

69 'Cámara de Diputados', *El Comercio*, 13 August 1845.

70 Yacobaccio, 'The Historical Relationship between People and the Vicuña', p. 17.

71 Emily Wakild, 'Saving the Vicuña: 'The Political, Biophysical and Cultural History of Wild Animal Conservation in Peru, 1964–2000', *The American Historical Review* 125:1 (2020), pp. 54–88.

72 Helen Cowie, *Victims of Fashion* (Cambridge: Cambridge University Press, 2022), pp. 148–167.

73 Philippa K. Wells, '"An Enemy of the Rabbit": The Social Context of Acclimatisation of an Immigrant Killer', *Environment and History* 12 (2006), pp. 297–324.

74 'Animals Verging on Extinction', *The Animal World*, February 1913, pp. 27–30.

75 Nature Notes: The Owl-Parrot', *Evening Post*, 3 October 1936.

76 'The Creatures Unlikely to Survive Another Year of Attrition', *The Times*, 3 January 1996.

77 'Rare Parrot Survives the Crush', *The Times*, 7 March 2014.

78 'Kākāpō Recovery', retrieved from www.doc.govt.nz/our-work/kakapo-recovery.

79 'World's Fattest Parrot, the Endangered Kākāpō, Could be Wiped Out by Fungal Infection', *The Guardian*, 13 June 2019.

80 Kakapo Quest', *Press Association*, 14 November 1989.

81 Rachel Thomas, 'Sirocco: The Celibate Spokesbird Worth $80,000' (27 June 2020), retrieved from www.rnz.co.nz/news/national/419984/sirocco-the-celibate-spokesbird-worth-80-000.

82 'Europeans Care More About Elephants than People, Says Botswana President', *The Guardian*, 17 April 2024.

Further Reading

Barrow, Mark, *Nature's Ghosts: Confronting Extinction from the Age of Jefferson to the Age of Ecology* (Chicago: University of Chicago Press, 2009)

Cowie, Helen, *Victims of Fashion: Animal Commodities in Victorian Britain* (Cambridge: Cambridge University Press, 2022)

De Bont, Raf, *Nature's Diplomats: Science, Internationalism and Preservation, 1920–1960* (Pittsburgh: University of Pittsburgh Press, 2021)

Dorsey, Kurk, *Whales and Nations: Environmental Diplomacy on the High Seas* (Seattle: University of Washington Press, 2013)

Gissibl, Bernhard, *The Nature of German Imperialism: Conservation and the Politics of Wildlife in Colonial East Africa* (New York: Berghahn, 2016)

Heise, Ursula, *Imagining Extinction: The Cultural Meanings of Endangered Species* (Chicago: University of Chicago Press, 2016)

Hennessy, Elizabeth, *On the Backs of Tortoises: Darwin, the Galápagos and the Fate of an Evolutionary Eden* (New Haven: Yale University Press, 2019)

Isenberg, Andrew, *The Destruction of the American Bison* (Cambridge: Cambridge University Press, 2001)

Jørgensen, Dolly, *Recovering Lost Species in the Modern Age: Histories of Longing and Belonging* (Cambridge, MA: MIT Press, 2019)

Kolbert, Elizabeth, *The Sixth Extinction: An Unnatural History* (London: Bloomsbury, 2014)

Krech III, Shepard, *The Ecological Indian* (New York: W.W. Norton, 1999)

Melville, Elinor, *A Plague of Sheep: Environmental Consequences of the Conquest of Mexico* (Cambridge: Cambridge University Press, 1994)

Miller, Shawn William, *An Environmental History of Latin America* (Cambridge: Cambridge University Press, 2007)

Paddle, Robert, *The Last Tasmanian Tiger: The History and Extinction of the Thylacine* (Cambridge: Cambridge University Press, 2000)

Quintero Toro, Camilo, *Birds of Empire, Birds of Nation* (Bogotá: Universidad de los Andes, 2012)

Soluri, John, *Creatures of Fashion: Animals, Global Markets and the Transformation of Patagonia* (Chapel Hill: Duke University Press, 2024)

Songster, E. Elena, *Panda Nation: The Construction and Conservation of China's Modern Icon* (Oxford: Oxford University Press, 2018)

Swart, Sandra, 'Zombie Zoology: History and Reanimating Extinct Animals', in Susan Nance (ed.), *The Historical Animal* (Syracuse: Syracuse University Press, 2015), pp. 54–74

Tucker Jones, Ryan, *Empire of Extinction* (Oxford: Oxford University Press, 2014)

Turvey, Nigel, *Cane Toads: A Tale of Sugar, Politics and Flawed Science* (Sydney: University of Sydney Press, 2013)

Van Dooren, Thom, *Flight Ways: Life and Loss at the Edge of Extinction* (New York: Columbia University Press, 2014)

Van Dooren, Thom, *A World in a Shell: Snail Stories for a Time of Extinctions* (Cambridge, MA: MIT Press, 2022)

Vasile, Monica, 'From Reintroduction to Rewilding: Autonomy, Agency and the Messy Liberation of the European Bison', *Environment and History* (2022), doi:10.3197/096734022X16552219786618

Wakild, Emily, 'Saving the Vicuña: 'The Political, Biophysical and Cultural History of Wild Animal Conservation in Peru, 1964–2000', *The American Historical Review* 125:1 (2020), pp. 54–88

Walker, Brett, *The Lost Wolves of Japan* (Seattle: University of Washington Press, 2005), pp. 129–83.

7   Animal Rights

Many human–animal relationships have involved an element of cruelty. Since their domestication thousands of years ago, animals have been subjected to overwork, captivity, brutal methods of training, selective breeding (often resulting in physical deformities), inhumane methods of slaughter and a range of scientific experiments. Some have also been victims of more gratuitous forms of cruelty, such as baiting, neglect or unprovoked teasing and abuse. In 1894, for instance, a man named George Swan visited the National Zoo, Washington DC, and 'squirted a mouthful of tobacco juice into the eye of a monkey to see how it would [re]act'.[1] Overall levels of animal suffering have almost certainly increased since the Industrial Revolution as more and more animals have experienced factory farming, commercial hunting and experimental procedures. At the same time, however, concern for animals has also increased significantly over the past two centuries, spawning a growing body of humanitarian organisations and animal welfare legislation.

This chapter traces shifting attitudes towards animal welfare. Previous chapters have explored the specific ethical issues posed by different forms of exploitation, from confinement in zoos to the suffering of animals in war. Here I focus on the rise of the animal protection movement since the early nineteenth century and explore how far this has changed the way we treat other species. To what extent have definitions of cruelty changed over time? What factors have shaped human perceptions of animal suffering? Has the fight for animal rights complemented, or worked against, the fight for the rights of disadvantaged humans? Where do animal welfare and conservation converge, and where do they diverge? The chapter begins with an overview of the emergence and expansion of the animal welfare movement, charting its shifting emphases and priorities. It goes on to explore some of the challenges and tensions within animal protectionism.

DOI: 10.4324/9781003181996-8

The Rise of the Animal Protection Movement

Animals have suffered mistreatment since the earliest periods of human history. Human concern for animal suffering, however, is a more recent phenomenon – at least in the West, and is usually dated back to the eighteenth century, when Enlightenment ideas and new social conditions combined to bring about a shift in human–animal relations. As we will see, this did not necessarily result in a reduction in actual cruelty to animals, but it did change the way in which that cruelty was perceived.

Before the eighteenth century, individual people undoubtedly cared for particular animals and may have treated them kindly. Some religions, moreover, encouraged respect for other species, and even, in the case of certain strains of Buddhism, vegetarianism. The Buddhist monk Zhuhong (1535–1615) condemned the wearing of silk (which involved the killing of silkworms) and the eating of meat.[2] The Dominican friar Martín de Porres (1579–1639) ran a hospital for animals at his sister's home in Lima, Peru, and nursed sick oxen at the Limatambo hacienda.[3] The Shogun Tokugawa Tsunayoshi of Japan issued 'clemency orders' from 1687 making it a crime 'to kill, abandon or torment animals of any sort, or to eat their flesh'. He also constructed large kennels in Edo for housing stray dogs.[4] These personal acts of kindness were not always sustained, however, and there was no coordinated movement to protect animals from cruelty. In early modern Europe, moreover, acts of extreme cruelty to animals were common, from dog slaughter to bear baiting, while Christian doctrine preached human dominion over other species. The seventeenth-century French philosopher René Descartes even argued that animals were automata, without rationality or sentience and incapable of feeling pain – though his ideas were not universally accepted by his contemporaries.

From the mid-1700s a combination of factors re-shaped human interactions with, and perceptions of animals – at least in northern Europe. On the one hand, new religious movements and intellectual trends prompted people to view animals with greater levels of sympathy and to question whether they should enjoy some of the rights increasingly being enshrined in law for humans. The Methodist John Wesley, for instance, advocated kindness to animals and treated his own horses with compassion. The philosopher Jeremy Bentham argued that animals were worthy of moral consideration on account of their sentience ('The question is not, can they *reason*, nor can they *talk*, but can they *suffer*?'). On the other hand, major social changes affected the way people perceived animals, making acts of cruelty more visible and rendering some longstanding traditions obsolete. Urbanisation, for instance, exposed growing numbers of middle-class residents to the sight of violence towards animals on city streets, as draught animals were driven with ever greater intensity and cattle were herded through urban streets prior to slaughter. Industrialisation, meanwhile, required workers to keep pace with the machine rather than setting their own working rhythms, and was incompatible (at least in employers' minds) with drawn-out and violent sports like bull baiting. Together, these political, intellectual and social changes made cruelty to animals both less acceptable and simultaneously more visible, sharpening calls for reform.[5]

The first concrete steps to protect (some) animals from abuse were taken in Britain at the turn of the nineteenth century. In 1800, following a rise in popular sentiment against blood sports, MP William Pulteney introduced a bill into Parliament to abolish bull baiting – a longstanding pastime in which dogs were set on a bull and attempted to grab onto its lips and nostrils with their teeth. This bill, and several similar ones, failed to secure enough votes to become law. In 1822, however, the MP for Galway, Richard Martin, succeeded in pushing an animal cruelty bill through Parliament that made it a crime to mistreat any farm or draught animal. This was followed by a second, more comprehensive, Cruelty to Animals Act in 1835, which abolished bull and bear baiting and extended protection to all domestic animals. To police the new legislation, a group of animal welfare advocates – including Martin – combined forces in 1824 to found the Society for the Prevention of Cruelty to Animals – the first organisation of its kind anywhere in the world. Initially based in London, the Society (which received royal patronage in 1840, becoming the Royal Society for the Prevention of Cruelty to Animals) quickly spawned branches across the country, providing national coverage for the fledgling animal protection movement.[6]

From Britain, the animal welfare movement gradually spread across the globe in the nineteenth century, with new legislation and animal protection organisations emerging in multiple countries. In 1850, the French Second Republic passed the Grammont Law, banning the public (though not the private) mistreatment of animals. In 1861 Colesworthy Grant – a British expat and artist – founded the first Society for the Prevention of Cruelty to Animals in India in Calcutta. In 1866 Henry Bergh founded the American Society for the Prevention of Cruelty to Animals in New York, taking the RSPCA as his model. In 1871 the first animal protection societies were established in Italy and Australia. As the locations of these organisations indicate, many nineteenth-century animal protection organisations were modelled closely on the RSPCA or directly set up by British (and later American) humanitarians in non-European settings. They did not, therefore, necessarily reflect the attitudes of local people and were, indeed, often promoted as part of a wider 'civilising' agenda. By the mid-twentieth century, however, as decolonisation swept the globe, an increasing number of former colonies began to enact legal protections for animals and to create their own home-grown organisations to enforce them. India, for instance, passed the Prevention of Cruelty to Animals Act in 1960 and set up the

Animal Welfare Board of India to administer it.[7] The first Chinese animal protection organisation, the Nanjing Society for the Prevention of Cruelty to Animals was founded in 1934, though lapsed during the Maoist era and was not revived until 1992, when the Chinese Small Animal Protection Society became the first animal welfare movement to attain recognition under the People's Republic of China (China still has no nationwide animal welfare law).[8] Bolivia enacted the Law for the Defense of Animals against Acts of Cruelty and Mistreatment in 2015 and became the first country in the world to ban all animal acts (wild and domestic) in circuses in 2009.[9]

What tactics have animal protection movements employed, and how have their priorities changed over time? If we start with tactics, these can be divided into three distinct approaches – often (but not always) pursued in parallel with one another: securing legal protection for non-human animals, educating the wider public about animal cruelty and taking direct action to alleviate the suffering of individual animals. In the case of the RSPCA, the organisation campaigned for changes to the law to ensure that all categories of animal were properly protected from abuse and funded a team of uniformed inspectors to patrol the streets enforcing the existing legislation. The latter investigated reports of cruelty and prosecuted individuals who had subjected animals to mistreatment; in 1849, the Society prosecuted a greengrocer named Tempest Fletcher for 'cutting and maiming a dog of the St Bernard breed' after it 'stopped in front of the defendant's shop for the purpose of nature'.[10] As well as targeting those who abused animals, the RSPCA funded multiple educational campaigns to improve the treatment of other species, running essay competitions for children, publishing a monthly magazine from 1869 and circulating pamphlets exposing various forms of animal cruelty, from the hunting of seals for their fur to the shipping of live cattle across the Atlantic. Since the twentieth century, the Society has also taken up the mantle of housing, treating and rehoming abandoned pets, building on the work of earlier organisations such as Battersea Dogs' Home (founded in 1860).

As for priorities, it is possible to identify key phases of concern as animal lovers responded to new or more prevalent threats to animal welfare. If we take the RSPCA once more as our primary example, the organisation focused its attention initially on the well-being of working animals (primarily horses and cattle) on British streets, which made up the majority of its convictions in its first decades of existence. It also prioritised the elimination of blood sports such as bull and bear baiting, which were seen as both cruel and socially corrosive. From the 1870s, vivisection and the use of animals in science became a new focus for the Society's campaigning, resulting in the revised Cruelty to Animals Act of 1876, which sought to regulate (though did not abolish) experimentation on animals. Later in the nineteenth century – and into the early twentieth – the emphasis shifted to the mistreatment of performing animals (culminating in the Performing Animals Act of 1925), the exploitation of a range of wild animals for fashion (especially fur and feathers), the protection of pit ponies (used in Britain's coal mines) and the hunting of animals for sport.[11] The RSPCA, a comparatively moderate organisation, did not advocate vegetarianism or a complete cessation of vivisection, but other more radical organisations soon began to do so, among them the Humanitarian League, which promoted a vegetarian diet, opposed keeping wild animals in captivity and supported a wide range of social reforms – including the abolition of the death penalty and female suffrage.[12] Moving into the twentieth century the division between welfare-based movements, like the RSPCA, and rights-based movements like People for the Ethical Treatment of Animals (PETA) has become wider, reflecting different aims and convictions. The rise of factory farming, moreover, and the massive expansion in the number of animals used globally in scientific experiments, have focused increasing attention on the cruelties associated with intensive agriculture and animal testing. The priorities of the animal protection movement have thus shifted over time in response to changing attitudes and social conditions.

Some Animals More Equal than Others?

Since 1800, there has been growing concern about cruelty to animals, beginning in Europe and spreading to the rest of the world. Not all forms of cruelty have, however, been regarded as equally serious, and not all have attracted equal levels of attention. Attitudes towards animal suffering are often culturally dependent and can change significantly over time. The RSPCA, for instance, advocated the culling of tigers in India in the 1870s, on the grounds that they 'cannot be civilised, but on the contrary are destined to destroy human life' – something it would never do today.[13] What, then, have been the key drivers in shaping human perceptions of animal suffering and what kinds of abuses have attracted most attention? Four factors stand out as being particularly important: the nature of the suffering, the purpose of the suffering, the visibility of the suffering and, crucially, the species of the victim.

The level of concern elicited by animal cruelty has been strongly influenced by its nature and duration. If an animal's suffering is relatively brief, or its pain alleviated by anaesthetic, criticism has tended to be limited. Protracted abuse, by contrast, has generated the most vociferous opposition, especially if it is repeated on more than one occasion. Commenting on the evils of vivisection, the prominent Victorian antivivisectionist Frances Power Cobbe argued that scientific experimentation was worse than hunting animals for sport, because the latter died quickly while the former experienced prolonged pain:

[I]t is almost ludicrous to compare a fox-hunt, for example, with its free chances of escape and its almost instantaneous termination in the annihilation of the poor fox when captured, with the slow, long drawn agonies of an affectionate, trustful dog, fastened down limb by limb and mangled on its torture trough.[14]

Terminating the life of an animal quickly and humanely was therefore acceptable – at least to some humanitarians – but long-drawn-out abuse was not.

As well as the nature of the suffering, its purpose was also important, with the key distinction here being between 'necessary' and 'unnecessary' cruelty. Such categories are, of course, fluid and open to interpretation. In general, however, the killing of animals for sustenance, for (essential) clothing or (more recently) to save or prolong human lives by advancing medical knowledge, have been classed as necessary usage, while inflicting pain on animals for entertainment, for sport or to satisfy the every-changing demands of fashion has been regarded as unnecessary or 'wanton' cruelty. This explains the early emphasis of the RSPCA on blood sports such as bull and bear baiting, which were deemed a form of gratuitous cruelty. It also explains the outrage expressed at animal experiments whose direct benefits to humans were unclear (such as Harry Harlow's controversial psychological experiments on monkeys in the 1950s), and at sartorial trends such as 'murderous millinery' at the turn of the nineteenth century, which saw millions of birds slaughtered to provide adornment for women's hats. Though not necessarily more painful than the use of animals for food or essential labour, the exploitation of other species for amusement, scientific curiosity or personal vanity has often been deemed more heinous because the ends have not justified the means. Why an animal is killed is often, therefore, as significant as how it is killed, even if the suffering, from the animal's perspective, is the same. Writing in 1875, for instance, one British lady condemned 'the dreadful cruelties' perpetrated against the Pacific fur seal to secure sealskin jackets for upper class women, which she regarded as a frivolous luxury, but accepted the use of seal oil, used widely in the nineteenth century for fuel and as a lubricant.[15]

A third factor that has had a major impact on perceptions of animal cruelty has been its level of visibility. Humans worry more about suffering that they can see than suffering that is hidden, so abuse that takes place on the streets, in the circus ring or at the zoo has tended to attract more censure than cruelty that happens in the laboratory, on the factory farm or in the wild. Early humanitarian movements, moreover, often explicitly connected animal cruelty with social disorder, fearing that those who witnessed it would become hardened to other forms of abuse. When a seal at the Manly Aquarium in Sydney feasted on a live shark before an audience of the paying public in 1887, the local Animal Protection Society complained about the 'indecent character' of the exhibition, which it feared might 'familiarise the growing generation with scenes which would have a tendency to brutalise the finer sensibilities'.[16] Aware that what is out of sight is often out of mind, those who profit from the exploitation of other species have frequently tried to conceal the more unpalatable aspects of their businesses, moving slaughterhouses to the edges of cities, outsourcing livestock production to colonial possessions or conducting painful experiments or training exercises behind closed doors. Humanitarian organisations, by contrast, have channelled much of their energy into exposing abuse and rendering the unseen visible, whether through emotive language and imagery or shocking undercover footage from laboratories, circuses and factory farms. In 1914, for instance, a Norfolk lady named Ada Cole oversaw the production of a film highlighting the cruelty inflicted on old horses exported from Britain to Belgium for slaughter, bringing a little-known trade into the public domain. As J. Keri Cronin has shown, the visual politics of animal welfare have played a crucial role in shaping perceptions of suffering, making seeing and bearing witness central elements of much humanitarian work.[17]

Lastly, species has played a significant role in shaping perceptions of, and responses to, animal abuse. Mammals and (to a lesser extent) birds have typically constituted the primary object of concern for humanitarians on account of their seeming proximity to human beings and their supposed higher levels of intelligence and capacity to feel pain. Within those groups, animals that interact closely with humans, correspond to human perceptions of cuteness and exhibit human-like emotions have invited particular claims for moral consideration, perhaps especially primates, which, at least from the twentieth century, have often been accorded the greatest capacity for suffering.[18] One big game hunter, for instance, recounted in 1924 how he shot a colobus monkey in Africa, after which the animal 'placed its paw over the wound and just looked at me, seeming to ask, "Why have you done it"?'. He resolved never to shoot a monkey again.[19] Another writer, Sir Hesketh Bell, observed captive orangutans at a dealership in the south of France and empathised with their sad plight, emphasising their human (and, indeed, canine) qualities: 'The orangs have the eyes of an Airedale, and one of them, near whom I stood for some time, looked into mine as if he were trying dumbly to tell me that his heart was breaking'.[20] While apes, elephants, dogs, cats and horses have thus elicited considerable sympathy from humans over the ages, reptiles, fish, amphibians and insects have generally attracted less concern, often on the assumption that they are less intelligent and feel pain less acutely. This has resulted in reduced public awareness of their suffering and their exclusion from some animal welfare legislation.[21] Invertebrates, such as locusts, crickets and mealworms, for instance, can still be fed as live prey to carnivores in British zoos, whereas vertebrates and cephalopods (octopuses and squid) cannot.[22] The US Farm Security and Rural Investment Act of 2002, meanwhile, amended the definition of 'animal' to exclude birds, rats and mice, permitting researchers to conduct experiments on them without legal restraints.[23] Species has therefore been a crucial determinant of rights and protections accorded to different animals, though cultural preferences have determined which species have been most revered in specific times and places.

To explore how these different factors operated in practice, let's look at a couple of case studies: the furore surrounding the treatment of an elephant named Gunda in an American zoo and the push in late nineteenth-century Britain to end the abuse of performing bears.

Gunda was a male Asian elephant at the Bronx Zoo. Imported in 1904 by zoo director William Temple Hornaday, Gunda was initially hugely popular, enjoying a brief career giving children elephant rides around the park. As he matured, however, Gunda's temper deteriorated and he became dangerous and unmanageable, causing headaches for his caretakers. In 1906, the elephant picked up a female visitor named Lucretia Hawes with his trunk, 'lifted her off her feet, and began to draw her over the brass railing toward the cage', piercing a vein in the back of her hand with his tusk.[24] A year later Gunda attacked one of his keepers, breaking several of his ribs.

Unable to contain the giant pachyderm, Hornaday resorted to tethering the animal in his den, chaining him by one fore- and one hindfoot to prevent him from turning on his keepers while they were cleaning his enclosure. This measure, however, elicited vociferous accusations of cruelty from the public, who expressed their concern in hundreds of letters to the *New York Times*. One critic pleaded, 'In the name of a decent civilization of even a limited degree of mercy and compassion, let this animal either be liberated or at once humanely destroyed'. Another meditated that 'whether Gunda is able to recall the images of freedom and rage dumbly over the reality of chains, or whether he is only conscious of the torment of inactive muscles and thwarted impulses, as he keeps up his wretched rocking and swaying on the only set of muscles allowed to act – in either case it is a pitiable sight, and the manifest cruelty of it is beyond argument'.[25] Angered by these allegations, Hornaday wrote a testy rebuttal, insisting that he knew what was best for Gunda and that the complaints of humanitarians were 'very likely to lead to the shooting of Gunda at an early date'.[26] Within a year, however, he, too, had become convinced that Gunda was suffering, enlisting the naturalist Carl Akeley to shoot the elephant in his enclosure – a decision he rationalised as putting the pachyderm out of his misery.

The popular outcry sparked by Gunda's alleged abuse highlights how the nature, duration and visibility of animal suffering influenced public responses. Chained in his enclosure for months at a time, Gunda's suffering was chronic and long lasting – something many observers found unacceptable. By contrast, Hornaday emphasised that Gunda's death was quick, thanks to the good marksmanship of his executioner, Akeley, which made it humane.[27] As a zoological exhibit, Gunda's plight was, of course, highly visible to the American public, which focused attention on his suffering – and, significantly, on the human impact of witnessing that suffering; one commentator remarked that visitors to the zoo 'must go away with a feeling of sadness instead of pleasure after seeing the living tomb of this tortured animal'.[28] The cruelty the elephant experienced was also, arguably, unnecessary, as his purpose was purely entertainment.

Perhaps most importantly, the fact that Gunda was an elephant was a big reason why he generated so much concern, for the species had a longstanding reputation for intelligence and sagacity. Often seen as sensitive and clever, elephants were frequently credited with human-like emotions and were believed to experience pain, loneliness and fear. Writing in 1900, one man expressed sorrow for the 'thirsty' female elephant at Lincoln Park Zoo, Chicago, describing how she 'opened a big, pink, pitiful throat' in search of water, 'stretching her trunk aloft and begging as plainly as a thirsty child'.[29] Two decades earlier, when a large African elephant named Jumbo was sold by London Zoo to American showman Phineas Taylor Barnum, a woman sent 'a parcel of crape and widow's weeds' to Alice, '(the female elephant in the next stall), [so] that she might mourn over her bereavement'.[30] Like his fellow pachyderms, Gunda was regarded as a sentient, emotional and highly intelligent animal – qualities that were all believed to accentuate his suffering. Such respect was not accorded to other, less iconic creatures, many of whom experienced similar, or worse, treatment than Gunda. As one contemporary observed: 'What a pity that those who lavish their sympathy on Gunda because his feet are tied with chains do not know of the real and needless suffering of cattle and hogs before being slaughtered, as they are suspended by the hind feet with a chain about two or three minutes before being "stuck" with the knife, all the time perfectly conscious and bellowing or squealing'.[31]

While the outcry over Gunda centred on one famous individual, the near-contemporary move against dancing bears reflected a wider shift in public opinion in relation to the training and exhibition of performing animals. Introduced to Europe from India by Romani travellers, dancing bears were a common sight on early modern streets and a popular form of entertainment (Figure 7.1).[32] From the mid-nineteenth century, however, animal lovers started to voice increasingly strong opposition to making bears dance in public, emphasising the cruelty inflicted on the performers, which typically had their claws and teeth filed and rings inserted in their noses. Critics alleged that the bears were beaten to make them perform, citing multiple examples of physical violence. In Britain, this led to a series of RSPCA prosecutions and several successful convictions for cruelty.

Figure 7.1 'A Performing Bear', *All About Animals* (London: George Newnes Ltd., 1898) p. 121.

Source: photograph by author.

Viewed against the criteria for concern outlined above, dancing bears, like Gunda, ticked several key boxes. First, the fact that much of the cruelty experienced by these animals happened in the streets made it highly visible, which explains why it was so frequently reported. People did not like to see animals being physically abused in public, and some denounced it; a crowd at Kingston-upon-Thames reportedly cried 'Shame' at a pair of German showmen who beat the paws of a blind bear 'with a stick until it cried out in its agony', indicating their disapproval.[33]

Second, as with Gunda, a particular issue with dancing bears was the prolonged, chronic nature of their pain, which several critics likened to torture. This was reflected strongly in newspaper reports about cruelty, which lingered over the details of the bears' injuries and emphasised the animals' distress. A bear abused in Wallingford in 1879 was described as 'flinch[ing] and shr[inking] when being led by [a] tight string' and 'occasionally groan[ing] with pain'.[34] The shoulder of a bear abused in Greenwich in 1882 was found to be a 'complete mass of weals' when it was examined by an RSPCA inspector – the result of repeated beatings.[35]

Third, and related to the long duration of the pain, animal welfare advocates worried increasingly about the behind-the-scenes cruelty inflicted on bears and other performing animals, which, they argued, far exceeded what was seen on the streets. In 1899, for instance, Joseph F. Simpson, 'Consulting Engineer to the Empire Theatre in Blackpool', visited the theatre in the daytime, when French showman Mr Permane was rehearsing for the evening's show, and witnessed a performing bear being beaten 'over the nose until its cries of agony could be heard all over the building'.[36]

Fourth, it is significant that the perpetrators of this abuse were often Frenchmen, Germans or Italians, which made it easier to present such cruelty as the preserve of foreigners. This once again highlights the cultural/racial dimension of animal protectionism in this period, and the degree to which kindness towards animals was perceived in Britain as a national trait. A contributor to *The Animal World* urged three German showmen to 'return to their Fatherland' after they were convicted for beating a 'blind and emaciated bear', remarking that, 'Whatever doubts may be entertained as to the alleged cruelty of the Germans in France [during the Franco-Prussian War], there can be no doubt whatever as to the cruelty of three Germans in England'.[37] Simpson, meanwhile, recounted how he went up to Mr Permane and 'told him in very plain English just what I thought of him', asserting that it is 'the right of every Englishman to interfere when he sees animals ill-used'.[38] The identity of the victim therefore mattered in cases of cruelty, but so, too, did the identity of the abuser – an issue to which we shall now turn.

Animal Rights versus Human Rights

Animal protectionism did not emerge in a vacuum. On the contrary, it has been heavily influenced by other reform movements, from abolitionism to feminism. It has also enjoyed a close, though sometimes conflicted, relationship with the nature conservation movement that emerged in Europe and the USA towards the end of the nineteenth century. To understand the wider significance and evolution of animal rights activism, therefore, we need to explore some of the connections between animal advocacy and other reform movements and to examine some of the fissures within the animal protection movement itself. What is the difference between a welfarist approach to other species and a rights-based approach to animals? Has the fight for animal protection complemented or worked against the fight for human equality? To what extent do animal conservation and animal protection operate in tandem? Where, how and why do they differ in emphasis?

If we begin by looking at the differences within the animal protection movement itself, we can identify two related but distinct traditions: a focus on animal welfare and a focus on animal rights. The first of these, a welfarist approach, concentrates on protecting animals from cruelty and minimising their suffering. Since 1979, this has often been associated with the 'five freedoms': namely freedom from hunger and thirst; freedom from discomfort; freedom from pain, injury or disease; freedom to express normal animal behaviour; and freedom from fear or distress.[39] Though welfarists recognise the value of animal life, an emphasis on welfare does not necessarily preclude the killing of animals – and, indeed, it often advocates euthanasia as a means of preventing further suffering (the 'five freedoms' were initially applied to the treatment of farm animals, many of which were destined to be slaughtered for their meat). It does, however, insist that death, where unavoidable, must be as painless as possible. Proponents of animal welfare therefore focus on alleviating suffering rather than completely transforming the human relationship with animals. This was the model followed by most nineteenth-century animal protection movements and it continues to be the priority of modern humane organisations like the RSPCA.

The animal rights movement (sometimes also called the animal liberation movement) concurs that reducing animal suffering is important, but it goes beyond that. Instead of arguing that current practices should be reformed to mitigate instances of cruelty, animal rights advocates call for an end to the rigid moral and legal distinction drawn between human and non-human animals, an end to the status of animals as property, and an end to their use in the research, food, fashion and entertainment industries. They believe that animals, like humans, possess certain inviolable rights, which should be enshrined in law. Tracing its origins back to the turn of the twentieth century, the animal rights movement came to the fore in the 1970s, following the publication of seminal philosophical texts such as Stanley Godlovitch, Rosalind Godlovitch and John Harris's *Animals, Men and Morals: An Inquiry into the Maltreatment of Non-humans* (1972), Peter Singer's *Animal Liberation* (1975) and Tom Regan's *The Case for Animal Rights* (1983). Since then, animal liberationists have campaigned to re-write the laws of different countries to accord greater rights to non-human animals, often using particular species or individuals as test cases. In 1994, for instance, a group of scientists, activists and conservationists launched the so-called Great Ape Project, which argued that our closest relatives – chimpanzees, bonobos, orangutans and gorillas – should enjoy basic human rights; specifically the right to life, the protection of individual liberty (which ruled out captivity in zoos) and freedom from torture (which ruled out scientific experimentation).[40] While animal welfarists focus on mitigating cruelty to animals, animal rights advocates thus call for the complete abolition of all forms of animal exploitation and a wholesale re-think of our relationship with other species.

How has the fight for animal rights intersected with (or diverged from) the fight for improved human rights? The picture here is mixed, showing a relationship marked by both cooperation and conflict. On the one hand, of course, it is possible to highlight many parallels between the fight for animal rights and the fight for human rights, which have often operated in tandem. In the early nineteenth century, for instance, calls to end bull baiting found support among prominent abolitionists such as William Wilberforce, while radical vegetarians like the Briton John Oswald explicitly linked the exploitation of animals for food with human enslavement, empire and monarchical oppression. At the turn of the twentieth century, members of the Humanitarian League campaigned not only for the better treatment of animals, but for an end to the death penalty and the introduction of female suffrage in Britain. In the twentieth and twenty-first centuries, meanwhile, campaigners for animal rights have frequently been at the forefront of efforts to protect indigenous peoples from exploitation, to advance civil and LGBTQ rights and to expose the abuses of abattoir workers (often some of the poorest members in society) as well as the animals they slaughter. Since the inception of the animal protection movement at the start of the nineteenth century, moreover, it is notable that women have played a prominent role in animal welfare organisations, making up the majority of the membership of the RSPCA and the ASPCA and founding the Royal Society for the Protection of Birds in 1889.[41] Anti-vivisectionists Frances Power Cobbe and Louise Lind-af-Hageby were both committed feminists, advocating for votes for women and drawing parallels between the non-consensual exploitation of animal bodies and the patriarchal efforts to control the bodies of human females.[42] All of this suggests that advancing the rights of animals has often gone hand-in-hand with fighting for the rights of marginalised human groups, from slaves to sex workers.

While advocacy for animals has thus often aligned closely with advocacy for disadvantaged social groups, there have also been instances when the protection of animals has come into conflict with the welfare of humans, or has been exploited to further socially conservative or imperialist agendas. First, if we look at the targets of animal welfare organisations, it is notable that most convictions for cruelty to animals have been of lower-class men and women, or, in the USA, of immigrant or non-white individuals, perpetuating class or race stereotypes. In nineteenth-century New York, for example, the ASPCA stigmatised Italian Americans as cruel for shooting songbirds and condemned kosher slaughter, which was practiced by Jewish communities.[43] In contemporary Peru the 1856 *Reglamento de Policía* prescribed different penalties for animal cruelty based on class, with slaves who 'seriously mistreated donkeys and other beasts of burden' receiving 'six strokes of the whip' and free people committing the same offence 'a fine or one or two pesos'.[44] On the other side of the Atlantic, the RSPCA denounced lower-class blood sports like bull baiting while turning a blind eye to elite sports like hunting and shooting, leading, in some cases, to accusations of hypocrisy. Prosecuted in 1913 for 'cruelly terrifying a lioness' by 'rattling the bars of [her] cage', travelling showman Albert Mander remarked, snidely, that 'I suppose there is no cruelty in a lot of dogs running after a fox' – a pastime still legally permitted in Britain at the beginning of the twentieth century.[45]

Second, as well as reflecting class biases, the fight against animal cruelty has fostered negative national stereotypes, and, in colonial settings, has served to criminalise longstanding cultural traditions. This has allowed Western (especially anglophone) nations to parade their moral superiority over their less 'enlightened' neighbours, using kindness towards animals as a marker of humanity. It has also permitted colonial authorities to assume the mantle of civilising power by abolishing cockfighting in early twentieth-century Cuba, protecting overloaded bullocks in nineteenth-century Calcutta or condemning horse sacrifice in 1950s Nigeria.[46] Even when criticising lapses in humanity in metropolitan contexts, moreover, animal protectionists have often taken the opportunity to congratulate themselves on their own superior sensitivity to animal suffering, contrasting their activism with the passivity of other, less caring, societies. Contributing to the public debate about Gunda in 1914, for instance, J. Edwin Russell, concluded that 'The spirited criticism of the treatment of ... the Indian elephant most strikingly shows that the American race is a humane and a feeling one' – a point he reinforced with reference to a chained bear he had seen in Cuba, whose sorry plight was 'smiled upon by onlookers'.[47]

Third, and more abstractly, there have been cases where animal suffering has appeared to take precedence over the suffering of human beings, raising concerns about misplaced priorities. Criticising the level of public concern for Gunda, one commentator complained that 'there are millions of human beings, not only in prisons, jails and asylums, but also in tenements and sweatshops, who have almost as little opportunity for exercise, fresh air and sunshine as the unfortunate beast'.[48] Why, then, the outpouring of sorrow for a bad-tempered elephant when so many human New Yorkers were living and working in grim conditions? More recently, campaigners for civil rights in the USA have expressed concern that the shooting of the famous lion Cecil by a trophy hunter in Zimbabwe in 2015 appeared to generate more public anger than the shooting of black citizens by the US police.[49] Efforts to protect animals from cruelty can, therefore, contribute to the fight against wider systems of oppression, but they can also reinforce existing social hierarchies and power relations, deflect attention from human suffering and threaten the liberty and livelihoods of lower-class or non-western people who rely on animals for their sustenance and labour.

If we turn, finally, to the relationship between animal welfare and animal conservation, we find a similar case of partial alignment and partial divergence, reflecting, once again, the differing emphases of the two movements.

On the positive side, of course, these movements have much in common, and animal welfare organisations have warmly supported conservation measures that not only help to preserve species (the primary goal of conservationists), but also reduce the suffering of individual animals (the priority of animal protectionists). Early twentieth-century conservation legislation that outlawed 'cruel' methods of hunting such as poisoning, trapping and killing animals in pitfalls all met with approval from humanitarians, who emphasised the protracted suffering caused by such techniques. Campaigns to save species like the egret and the fur seal from extinction by reducing consumer demand for their feathers and fur likewise saw conservationists and humanitarians in agreement – though the former focused primarily on population decline while the latter emphasised the suffering inflicted on egret chicks, left to starve after their parents were shot, or seals allegedly skinned alive to provide women with sealskin jackets.[50] Both humanitarians and conservationists also oppose factory farming, though again with differing emphases. For the former, the primary concern is the suffering of the farmed livestock, be they battery hens, pigs in farrowing crates or beef cattle on cramped feedlots. For the latter, the focus is on the impact of intensive farming on the environment, from pollution caused by pesticide runoff to the emission of methane by factory-farmed cattle.

While conservationists and animal welfare advocates generally share the same broad goals for non-human animals, however, there are areas in which the two disagree. In almost all cases, these relate to instances in which measures to conserve a species can be seen to impact negatively on the welfare of individual animals – something many humanitarians find unacceptable. If we take the issue of invasive species, for instance, there are obvious tensions between conservationists, who advocate culling the invaders, and humanitarians, who often object to this practice on the grounds of cruelty – or at least express concern as to the specific control methods used. A balance must thus be struck between the rights of native species and the rights of introduced species – all of which are sentient beings capable of suffering. If we consider, on the other hand, practices such as captive breeding and animal reintroductions (see Chapter 6), similar tensions exist between the needs of the species and the welfare of individual animals. For conservationists, it is necessary – if regrettable – to risk or compromise the lives of individuals in uncertain reintroduction projects to ensure the continued existence of the species in the wild. For animal welfare advocates, conversely, the rights of the individual also matter, and their sacrifice for a bigger goal is more questionable. Finally, and perhaps most contentiously, conservationists and welfare advocates differ sharply on the issue of trophy hunting, in which ranch owners in Africa make money by allowing wealthy foreigners to come and shoot the animals on the land for sport. From a conservation perspective, the practice, though distasteful, can be seen as beneficial to wildlife, as without this financial incentive, these lands would probably be turned over to agriculture, and crucial animal habitat lost. From a humanitarian perspective, however, trophy hunting is repugnant, and many animal rights groups have called for it to be banned. Conservationists and humanitarians are therefore frequent allies in the fight to secure the survival and ensure the well-being of non-human animals, but there are instances in which these goals are incompatible.

The Ethics of Animal Protection

To explore the complex ethics of animal protection, the chapter concludes with four case studies, each of which highlights the challenges of balancing animal rights against human rights and balancing the competing rights of different species. What rights should animals possess, and what does the granting of those rights mean for marginalised human groups? Is it acceptable to compromise the well-being of individual animals to preserve a threatened species? Do the rights of native species trump those of invasive species, or do all animals have an equal right to life – or, at the very least, to a humane death?

Rights for Apes

Apes share 99% of their DNA with humans. Despite this, humans have hunted them for meat, museum specimens and the exotic pet trade, kept them captive in zoos and circuses and used them as human surrogates in medical research and even space travel. This raises important questions about the legal rights of apes – and, by extension, of other species. Should chimpanzees, bonobos, orangutans and gorillas be granted some of the rights that humans enjoy, on account of their close biological relationship to humans? If so, which rights should they possess, and what would the possession of such rights mean in practice?

Calls for apes to be recognised legally as persons rather than property have generally been rejected in court. In 2015, however, in a landmark judicial decision, the Argentine judge, Elena Liberatori, ruled that a 29-year-old female orangutan named Sandra was not an animal, but a 'non-human person', and could no longer be kept in captivity. The ruling came in response to a test-case brought by an animal rights group, who argued that Sandra, a long-term resident of Buenos Aires Zoo, had been illegally detained, and should be set free. Born in captivity, Sandra could not be immediately released from the Zoo, as she would not have survived in the wild. In 2016, however, following the closure of Buenos Aires Zoo, the orangutan was sent to the Center for Great Apes in Wauchula, Florida, where she learned to forage for her food, befriended a male orangutan named Jethro and became an internet star after she was filmed washing her hands with soap and water.[51]

Sandra's case represents a major victory for the animal rights movement, giving legal sanction to the notion that great apes should enjoy key human rights. It also raises complex ethical questions about our relationship to other species and how it should be framed. Is it right, for instance, to treat great apes with greater moral consideration on account of their proximity to humans, and, if so, what legal rights should they possess? If we decide that apes like Sandra merit greater protection than other species, what exactly are our criteria for granting these rights? Is it the animals' level of cognition or their apparent capacity for suffering? In either case, what is the justification for drawing the line at great apes? Elephants, pigs, dogs, dolphins and crows have all exhibited several key markers of intelligence (judged according to human standards), so why should such rights not also extend to them?[52] Approaching the issue from the opposite direction, does the granting of rights to apes erode the rights of other human groups – for instance infants, or those with learning disabilities, who might not, in some cases, meet the same markers of sentience as great apes? And is it even morally right to worry about the mental state of a single orangutan when millions of humans are starving, homeless or perishing in war? Do legal cases like Sandra's, moreover, bring tangible benefits for animals, or are they a distraction from efforts to alleviate the suffering of much larger numbers of non-human creatures? Sandra herself is undoubtedly living in better conditions in Wauchula than she was in Buenos Aires Zoo, but she is still, ultimately, in captivity. What does it mean, then, in practice, to grant legal rights to animals when humans have transformed their lives and (in many cases) their natural habitats beyond recognition?

Pups to the Slaughter

Since 2010 the Chinese city of Yulin in Guanxi Province has played host to a grisly festival. Every year, during the summer solstice, around 10,000 dogs are shipped into the city and clubbed to death so that their meat can be consumed by humans in local restaurants. The dogs are brought to Yulin from across China and kept in appalling conditions. Cats are also killed for human consumption, though their numbers are harder to calculate. Though presented by the authorities as a traditional event and a boost to local tourism, the Yulin festival has elicited vociferous opposition from local and international animal activists who have called for the slaughter to be banned. In 2015, 3.8 million people globally signed petitions against the festival, making it something of a cause célèbre for the animal protection movement.

The strong emotions generated by the Yulin festival can be attributed to three primary factors: the status and species of the victims, the dangers the practice may pose to human health and the visibility of the slaughter. First, it appears that many of the dogs killed for their meat are stolen. There are no official dog farms in China (unlike in Korea, where the consumption of dogmeat is more widespread), so most of the animals are taken from the streets, some of them strays, but others beloved pets. In 2016, for instance, when animal activists forced a lorry carrying dogs off the road in Hebei Province, one lady rushed to the scene and discovered her missing golden retriever inside 'amid hundreds of other whimpering dogs, many in poor condition'.[53] This is, of course, upsetting to many people, both in China, where the practice of pet-keeping has grown rapidly since the launch of China's economic reform in 1978, and in the West. Second, the consumption of dog meat raises concerns about rabies transmission and food safety. Many dogs are sick or infected and some have died through poisoning. This poses a threat to human health. Third, the dogs at Yulin are slaughtered in public, causing distress to urban viewers, who are increasingly distanced from such practices in their daily lives. They are also slaughtered in view of one another, increasing their terror and suffering. Critics argue that the sight of slaughter will desensitise the population to cruelty and corrupt children. For all these reasons, dog meat consumption violates food, welfare and public decency taboos and attracts global attention.

While the Yulin festival is undoubtedly cruel, the widespread hostility towards it raises its own questions and moral inconsistencies and highlights some of the ethical quandaries discussed earlier. On the one hand, supporters of the dogmeat industry argue that it provides employment for unskilled rural workers and benefits the economy by promoting tourism. Dogmeat consumption is highest in China's poorer provinces, so prohibiting it would affect them most (though, as Peter Li has shown, the actual number of people involved is 'negligible'). On the other hand, some of the international reaction to Yulin has racist overtones and fosters negative stereotypes of the Chinese as barbarous and cruel towards animals. This is, of course, problematic, and plays into older ideas about the link between kindness to animals and civilisation. It is also misrepresentative, given that most Chinese people never eat dogmeat and that (contrary to the claims of its supporters) the Yulin festival is a largely invented tradition (the consumption of dogmeat was banned under the Song dynasty (960–1279) and uncommon under the Qing dynasty (1644–1911), when dogs were kept as pets in Beijing's Forbidden City). Finally, as those who eat dogmeat point out, westerners eat beef and other meats, while in China vast amounts of pork are consumed – often at the cost of considerable cruelty to pigs. We know that pigs are as intelligent as dogs, so why does the plight of the former garner so much more attention? Is it right to demonise one form of cruelty while overlooking another?[54]

Violent Care

As we saw in Chapter 6, human actions have pushed many species to the brink of extinction. This, in turn, has led other humans to take drastic actions to ensure the survival of beloved or iconic species, breeding surviving individuals in captivity, using techniques such as surrogacy and artificial insemination to increase the number of offspring they produce and releasing animals into the wild to recolonise old habitats. Though done for the benefit of the species, all these actions involve elements of suffering for individuals of the target species and may compromise the welfare of individuals of other species as well. How far, then, should we go to save species from extinction, and which should take precedence: the welfare of the individual or the survival of the species?

One case that illustrates this dilemma especially powerfully is that of the whooping crane. The tallest bird in North America, and named for the 'whooping' call it makes to warn its partner of potential danger, the whooping crane once existed across the mid-west of the continent, stretching as far south as Mexico. By the 1930s, however, wetland loss and over-hunting had caused the birds' numbers to plummet, with only a single flock of around twenty birds migrating each year between Canada and Texas. Concerned that this tiny flock could be wiped out by one tragic accident, conservationists sought to create a second migrating population in the east of the US, travelling between Wisconsin and Florida. This was not a simple enterprise, but involved taking eggs from nesting cranes in Canada, using incubation and artificial insemination to increase reproduction rates, raising the resulting crane chicks in captivity with human carers (the latter dressed in crane costumes to prevent imprinting) and teaching adult cranes to fly south behind an ultralight aircraft – all incredibly challenging tasks. By 2020, thanks to these endeavours, there were an estimated 506 cranes living in the remnant original migratory population, 162 in three reintroduced populations and 144 birds in captivity, putting the total current population at just over 800.[55]

While the huge efforts made by conservationists have contributed to the survival of a highly endangered species, the operation to save the whooping crane has come at a heavy cost to individual birds, who have endured a degree of suffering, or at least discomfort and deprivation. Whooping crane chicks, for instance, have been raised by humans rather than by their parents and not all of them have been released into the wild – some have lived out their entire lives in captivity. Adult whooping cranes have forgone their liberty and undergone repeated artificial insemination procedures to maximise population growth and genetic diversity among chicks, a process that, despite the best efforts of carers, is likely to have been stressful and uncomfortable. Adults of other related species, meanwhile (such as sandhill cranes, chickens and geese), have been used as surrogate parents for whooping crane chicks, and, later, as test subjects for guided migration in place of the less expendable whooping crane adolescents, often succumbing en route to hunters, bad weather or ensnarement in power lines. The first whooping crane to be bred in captivity, Dawn, for instance, was raised in a pen alongside five young turkeys 'to show the young whooping crane how to eat, keep it company and prevent it becoming too attached to people'.[56] A group of seven sandhill cranes was used for the first aircraft-guided migration from Idaho to New Mexico, two of whom perished in attacks by golden eagles, and one of whom 'finished the trip in a van because it couldn't keep up'.[57] Though enacted with the best of intentions, therefore, measures to preserve the whooping crane as a species have consistently overridden the welfare of individual birds – a process the philosopher Thom Van Dooren characterises as 'violent care'.[58] Similar compromises have been made in other captive breeding and reintroduction programmes, from the release of golden lion tamarins in Brazil to the treatment of captive-bred ruffed lemurs in Madagascar (see Chapter 6).

Grey Squirrels and Narco Hippos

If efforts to save endangered species have come at a cost from a welfare perspective, what about the issue of invasive species? How do we balance the rights of introduced species against the rights of native species affected by their introduction? Is it ethical to cull the invaders, and, if so, what methods of slaughter are appropriate? Have all invasive species elicited the same responses from humans, or do some attract more concern than others? To explore this complex issue, let us consider two very different, but equally polarising, invasive species: the grey squirrel in Britain and the hippopotamus in Colombia.

Grey squirrels originated from North America, but were released at various sites in Britain between 1890 and 1920, quickly colonising large areas of the country. Initially seen as picturesque additions to the British countryside, they soon came to be viewed as a pest, owing to their depredations on gardens, trees and farms, their impact on native birds and the threat they were believed to pose to the native red squirrel, which was simultaneously in decline.[59] Writing in 1921, for instance, Stanley B. Hodgson expressed concern that 'further encouragement of this introduced species ... will inevitably lead, sooner or later, to the extermination of our own red squirrel'.[60] Twenty-five years later Edward Cadogan complained that 'In an area of about 50 acres [in the Chiltern Hills] the grey squirrel has completely destroyed every young sycamore tree and even some of a goodly size' by stripping away the bark.[61]

While scientists largely agreed that grey squirrels should be removed, efforts to control the species have elicited mixed reactions from the British public, many of whom found the squirrel pretty and charming. Rallying to the support of the rodent, for example, one early commentator, Maud Meugens, characterised the grey squirrel as 'sleek, well-groomed and exquisitely clean, and the most graceful of acrobats'; 'Next to puppies, they have the most engaging ways of any members of the animal kingdom I have ever seen'.[62] Another writer, F.C. Streatfield, related a grisly story about two carrion crows in Kensington Gardens 'doing to death' a young mistle-thrush and concluded that the corvids contributed more to the decline in native wild birds than grey squirrels. 'Besides, the squirrels are charming pets, and will sit on one's knee and eat nuts, while the crows are pure vermin'.[63] A third writer, Hugh F. Marriott, asked:

How many cats are there that do not revel in a succulent young bird, their natural prey, and who in nature has the greater right to the white heart cherries, the squirrel, who cannot be expected to differentiate between common and private property, or the human-being who has the produce of the world at his command?[64]

As these responses make clear, the squirrels' cuteness elicited doubts among some Britons as to the morality of exterminating them and prompted favourable comparisons with other, less desirable species. Even some of those who supported culling, moreover, acknowledged that killing such an attractive creature went against man's natural instincts, but must be done – humanely – as a matter of necessity. Writing in 1937, for instance, a *Times* journalist conceded that the grey squirrel was not one of those creatures 'which man, speaking generally, has an instinct to destroy', but asserted that 'To be kind to the grey squirrel is to be cruel to much else'.[65] Later the same year, Captain Charles Westley Hume, Hon. Secretary of the University of London Animal Welfare Society reminded his compatriots that 'the anti-grey squirrel campaign recommend a humane cage trap … for catching grey squirrels' and that 'in view of the law, it is not permissible to set gin traps for them on any pole, tree or cairn'.[66] Issues of animal welfare have thus featured prominently in discussions about the grey squirrel, delaying and limiting eradication efforts.

Hippopotami are native to Africa, but since the early 1990s, they have become an unusual fixture in Colombia's longest waterway. Illegally imported into South America by the notorious drugs lord, Pablo Escobar, an initial group of four hippos escaped from a private zoo at Hacienda Nápoles following their owner's death in 1993 and quickly colonised the nearby Magdalena River and its tributaries. With plentiful food, no native predators and – unlike in Africa – no seasonal droughts to keep the population in check, their numbers rapidly multiplied, rising to between 65 and 80 by 2020. The presence of hippos in Colombia is now having a measurable effect on the local environment, where they compete with native wildlife such as otters and manatees and pollute rivers and lakes with their excrement, decreasing oxygen levels in the water.[67] They also pose a potential danger to local people; one man suffered serious injuries in 2020 after he was attacked by a female hippo while fishing.[68]

As in the case of grey squirrels in Britain, it is generally accepted that Colombia's immigrant hippos are a problem. Once again, however, efforts to remove the animals have run into difficulties – some logistical and some ethical. On the one hand, the obvious solution – a cull of the pachyderms – has triggered opposition from animal rights organisations, who argue that this contravenes the rights and welfare of the hippos. One male hippo – Pepe – was shot in 2009, but his killing generated a public outcry, preventing a wider cull and sparing the life of his mate, Matilda.[69] On the other hand, non-lethal methods of population control such as sterilisation and returning the hippos into captivity have proven difficult to implement, partly for practical reasons and partly because they cost a lot of money. A scheme to castrate the male hippos, for example, has had some initial success, but is not progressing fast enough to slow reproduction.[70] A proposal to round up the hippos from the Magdalena and send them to zoos has likewise had limited impact, owing to the difficulty and expense of catching and transporting such bulky animals. Putting hippos in zoos, moreover, poses its own ethical quandaries – is it right to put what are now, essentially, wild animals back in captivity? With no universally acceptable solution on the horizon, the hippos look set to stay and to multiply, their population potentially rising to 1,500 by 2040.[71]

The cases of the grey squirrel and the narco hippo illustrate the challenge of dealing with invasive species and the hard choices that have to be made between individuals and ecosystems – especially when the former are cute or charismatic. The squirrels and the hippos present significant ecological problems in Britain and Colombia respectively, and are threatening biodiversity in both countries (though the precise impact of both species continues to be contested). Animal rights activists, however, object to euthanasia as a remedy, arguing that it would be cruel to the introduced species – who, after all, did not choose to be there. Many local people also support this position, having become fond of the invaders, having benefited economically from their appeal to tourists (in the case of the hippos) and, in some cases, having begun to treat them as pets. Squirrel-lover Maud Meugens, for instance, described how the squirrels living in her garden would 'come running up when we call, like tame pets, then sit up and beg, clasping their comfortable little persons with horny hands, living notes of interrogation.[72] One Colombian woman, meanwhile, reported that her family adopted a young hippo called Luna, feeding her regularly with milk.[73] Dealing with the invasive squirrels and hippos therefore means weighing the environmental equilibria of the British woodlands and the Magdalena River against the well-being of the introduced species. It also means balancing the welfare of the grey squirrels/hippos against the welfare of native species, who may also suffer if the latter remain at large. Should the well-being of grey squirrels take precedence over the well-being of red squirrels or nesting birds in Britain's woodlands? Should the lives of hippos take precedence over the survival of native fish, otters and manatees in Colombia's rivers? Who should make these decisions: scientists, local people or animal rights activists?

Conclusion

Cruelty has always been a feature of human–animal relations. From baiting bears in medieval England to extracting bear bile in twentieth-century China, humans have subjected other species to pain, loss of liberty and death. Since the mid-nineteenth century, animal suffering has generally become less visible to most humans, as abattoirs have been moved to the edge of cities and working animals have disappeared from city streets. While the visibility of animal cruelty may have decreased, however, its prevalence has almost certainly increased, with growing numbers of animals being reared for meat on factory farms, experimented on in science research facilities and hunted for their skins, horns or ivory. We may no longer see bulls being baited in London or captive giraffes being paraded through Florence, but our rising demand for meat has brought misery to millions of chickens, cattle and pigs, who spend most of their short lives in confinement.

Although the net suffering of animals has risen since the nineteenth century, human compassion for other species has also grown over the same period, as people have become more aware of animal sentience. This has given rise to a global animal welfare movement, beginning in Britain in the 1820s and spreading to the rest of the world. While animal welfare charities have made substantial strides in alleviating animal suffering, animal protection poses challenging ethical questions and has often entailed difficult trade-offs between the rights and survival of different animals. Should pet cats' right to freedom trump wild birds' right to life? Should Colombia's hippos be culled or castrated to protect the country's indigenous fauna? Should sandhill cranes be sacrificed to save whooping cranes from extinction? Should apes enjoy rights denied to elephants? In tackling these complex problems, animal protectionists must make controversial and often difficult decisions, balancing the competing rights of different species – including humans.

Notes

1 'Fined for Cruelty to a Monkey', *Washington Post*, 15 March 1894.
2 Joanna Handlin-Smith, 'Liberating Animals in Ming-Qing China: Buddhist Inspiration and Elite Imagination', *Journal for Asian Studies* 58:1 (1999), pp. 57–58.

3 Abel Alves, *The Animals of Spain* (Leiden: Brill, 2011), pp. 169–182.
4 Brett Walker, *The Lost Wolves of Japan* (Seattle: University of Washington Press, 2005), pp. 80–83.
5 Hilda Kean, *Animal Rights: Political and Social Change in Britain since 1800* (London: Reaktion Books, 1998), pp. 13–38.
6 Kathryn Shevelow, *For the Love of Animals: The Rise of the Animal Protection Movement* (New York: Henry Holt, 2008), pp. 201–222.
7 Andrew Linzey (ed.), *The Global Guide to Animal Protection* (Urbana: University of Illinois Press, 2013), pp. 9–35.
8 Peter Li, *Animal Welfare in China* (Sydney: University of Sydney Press, 2021), pp. 267–310.
9 Linzey (ed.), *The Global Guide to Animal Protection*, p. 257.
10 'Clerkenwell', *The Standard*, 11 June 1849.
11 Brian Harrison, 'Animals and the State in Nineteenth-Century England', *English Historical Review* 88 (1973), pp. 786–820.
12 Dan Weinbren, 'Against All Cruelty: The Humanitarian League, 1891–1919', *History Workshop* 38 (1994), pp. 86–105.
13 'Preservation of Tigers', *The Animal World*, May 1876, pp. 69–70.
14 Frances Power Cobbe, 'The Moral Aspects of Vivisection', in *The Modern Rack* (London, 1889), p. 10.
15 'Cruelty to Seals', *The Animal World*, April 1875, p. 62.
16 'The Recent Shark and Seal Fight', *Daily Telegraph*, 16 November 1887.
17 J. Keri Cronin, *Art for Animals: Visual Culture and Animal Advocacy, 1870–1914* (University Park, PA: The Pennsylvania State University Press, 2018), pp. 87–92.
18 On the role of physical proximity, bodily likeness, conceptions of sentience and modes of interaction in shaping human responses to animals, see Steven Wagschal, *Minding Animals in the Old and New Worlds: A Cognitive Historical Analysis* (Toronto: University of Toronto Press, 2018); and Helen Cowie, '"Down Pythons' Throats We Thrust Live Goats": Snakes, Zoos and Animal Welfare in Nineteenth-Century Britain', *British Journal for the History of Science* (2024), https://doi.org/10.1017/S0007087424000542.
19 'Gorilla Hunting', *The Times*, 19 June 1924.
20 'Prisoners and Captives: Orang-utans from Sumatra', *The Times*, 15 May 1928.
21 John Simons, *Goldfish in the Parlour* (Sydney: University of Sydney Press, 2023), pp. 265–272.
22 Geoff Hosey, Vicky Melfi and Sheila Pankhurst, *Zoo Animals: Behaviour, Management and Welfare* (Oxford: Oxford University Press, 2013), pp. 434–435.
23 Marc Bekoff, *The Emotional Lives of Animals* (Novato: New World Library, 2007), p. 139.
24 'Elephant Seizes Woman', *Chicago Daily Tribune*, 12 August 1906.
25 'Times Readers Protest Against Gunda's Imprisonment', *New York Times*, 19 July 1914.
26 'Hornaday on Gunda', *New York Times*, 27 June 1914.
27 'The Rights of Wild Animals', *New York Tribune*, 24 September 1922.
28 'Times Readers Protest Against Gunda's Imprisonment', *New York Times*, 19 July 1914.
29 'Says Elephant is Thirsty', *Chicago Daily Tribune*, 27 May 1900.
30 'Jumbo', *The Animal World*, March 1884, p. 34.
31 'Times Readers Protest Against Gunda's Imprisonment', *New York Times*, 19 July 1914. For a detailed appraisal of Gunda's case, see Nigel Rothfels, *Elephant*

Trails: A History of Animals and Cultures (Baltimore: Johns Hopkins University Press, 2022), pp. 85–123.

32 Deniz Dölek-Sever, 'Captive Wild Animals as Visual Commodities in the Ottoman Empire: A Historical Review', *Middle Eastern Studies* (2023), doi:10.1 080/00263206.2023.2215716.

33 'Cruelty to Ferae Naturae', *The Animal World*, 1 December 1870, p. 40.

34 'Cruelty to a Performing Bear', *Portsmouth Evening News*, 16 May 1879.

35 'Cruelty to a Performing Bear', *Manchester Courier*, 11 April 1882.

36 'Almighty Man and his Prey, No.2: The Training of Stage Bears', *The Animals' Friend*, vol. 5 (1899), p. 139.

37 'Cruelty to Ferae Naturae', *The Animal World*, December 1870, p. 40.

38 'Almighty Man and his Prey, No.2: The Training of Stage Bears', *The Animals' Friend*, vol. 5 (1899), p. 139.

39 Bekoff, *The Emotional Lives of Animals*, p. 163.

40 Paola Cavalieri and Peter Singer (eds), *The Great Ape Project: Equality Beyond Humanity* (New York, St. Martin's Press, 1994), pp. 4–7.

41 Diana Donald, *Women against Cruelty: Protection of Animals in Nineteenth-Century Britain* (Manchester: Manchester University Press, 2019).

42 Coral Lansbury, *The Old Brown Dog: Women, Workers and Vivisection in Edwardian England* (Madison: University of Wisconsin Press, 1985).

43 Janet Davis, *The Gospel of Kindness: Animal Welfare and the Making of Modern America* (Oxford: Oxford University Press, 2016), pp. 84–115.

44 *Reglamento de Policía para la Capital de la República y su Provincia* (Lima: Imprenta de Eusebio Aranda, 1856), pp. 36–37. My thanks to Steven Navarrete Cardona for this reference.

45 'Terrified Lioness', *Manchester Courier*, 3 January 1913.

46 Davis, *The Gospel of Kindness*, pp. 116–150; Samiparna Samanta, *Meat, Mercy and Morality: Animals and Humanitarianism in Colonial Bengal, 1850–1920* (Oxford: Oxford University Press, 2021), pp. 206–242; Saheed Aderinto, *Animality and Colonial Subjecthood in Africa* (Athens: Ohio University Press, 2022), pp. 227–248.

47 'Animal Cruelty in Cuba', *New York Times*, 2 July 1914.

48 'Times Readers Protest Against Gunda's Imprisonment', *New York Times*, 19 July 1914.

49 Bénédicte Boisseron, *Afro-Dog: Blackness and the Animal Question* (New York: Columbia University Press, 2018), pp. 32–35.

50 Helen Cowie, *Victims of Fashion* (Cambridge: Cambridge University Press, 2021), pp. 17–86.

51 'Orangutan Really Is Just Like You Ooh Oooh, Court Rules', *The Times*, 15 May 2015; 'This Orangutan's "Personhood" Victory Brings Hope to US Animal Rights Movement', *The World*, 20 November 2019.

52 'African elephants Pupy and Kuky remain on display at Buenos Aires Zoo (now converted into an Eco Park), despite meeting many of the criteria for 'personhood' ascribed to Sandra.

53 'Chinese Dogs Saved from the Stir Fry by Activists', *The Times*, 23 November 2016.

54 For a detailed appraisal of dog meat consumption in China, see Li, *Animal Welfare in China*, pp. 109–148.

55 Wade Harrell and Mark Bidwell, *Report on Whooping Crane Recovery Activities: 2019 Breeding Season – 2020 Spring Migration* (Canadian Wildlife Service and US Fish & Wildlife Service, 2020), p. 1.

56 'Dawn Heralds New Day for Whooping Cranes', *Washington Post*, 1 June 1975.

57 'Flying Away Home with "Mother"', *Washington Post*, 10 November 1996.

58 Thom Van Dooren, *Flight Ways: Life and Loss at the Edge of Extinction* (New York: Columbia University Press, 2014), pp. 87–122.
59 Peter Coates, 'A Tale of Two Squirrels: A British Case Study of the Sociocultural Dimensions of Debates over Invasive Species', in Reuben Keller, Marc Cadotte and Glenn Sandiford, *Invasive Species in a Globalised World: Ecological, Social and Legal Perspectives on Policy* (Chicago: University of Chicago Press, 2015), pp. 44–71.
60 'The Grey Squirrel', *The Times*, 23 December 1921.
61 'The Grey Squirrel', *The Times*, 26 August 1936.
62 'The Grey Squirrel', *The Times*, 21 December 1921.
63 'A Tragedy of Bird Life', *The Times*, 12 May 1924.
64 'Grey Squirrels', *The Times*, 23 August 1937.
65 'The Grey Squirrel', *The Times*, 3 July 1937.
66 'Grey Squirrels', *The Times*, 24 July 1937.
67 Jonathan Shurim et al., 'Ecosystem Effects of the World's Largest Invasive Animal', *Ecology* 101:5 (2020), pp. 1–9.
68 'Colombia define la suerte de los hipopótamos de Pablo Escobar', *El Comercio*, 21 March 2021.
69 'El Hipopótamo "Pepe" divide Colombia', retrieved from www.publico.es/actualidad/hipopotamo-pepe-divide-colombia.html.
70 'Colombia define la suerte de los hipopótamos de Pablo Escobar', *El Comercio*, 21 March 2021.
71 'Invasion of the Hippos', *Washington Post*, 11 January 2021.
72 'The Grey Squirrel', *The Times*, 21 December 1921.
73 'Colombian Drug Lord's Hippos Wallow in his Legacy', *The Times*, 28 June 2014.

Further Reading

Aderinto, Saheed, *Animality and Colonial Subjecthood in Africa* (Athens: Ohio University Press, 2022)
Alves, Abel, *The Animals of Spain* (Leiden: Brill, 2011)
Bekoff, Marc, *The Emotional Lives of Animals* (Novato: New World Library, 2007)
Boisseron, Bénédicte, *Afro-Dog: Blackness and the Animal Question* (New York: Columbia University Press, 2018)
Cavalieri, Paola, and Singer, Peter, *The Great Ape Project: Equality Beyond Humanity* (New York: St Martin's Press, 1993)
Cowie, Helen, '"Down Pythons' Throats We Thrust Live Goats": Snakes, Zoos and Animal Welfare in Nineteenth-Century Britain', *British Journal for the History of Science* (2024), https://doi.org/10.1017/S0007087424000542.
Cronin, J. Keri, *Art for Animals: Visual Culture and Animal Advocacy, 1870–1914* (University Park, PA: The Pennsylvania State University Press, 2018)
Davis, Janet, *The Gospel of Kindness: Animal Welfare and the Making of Modern America* (Oxford: Oxford University Press, 2016)
Donald, Diana, *Women against Cruelty: Protection of Animals in Nineteenth-Century Britain* (Manchester: Manchester University Press, 2019)
Handlin-Smith, Joanna, 'Liberating Animals in Ming-Qing China: Buddhist Inspiration and Elite Imagination', *Journal for Asian Studies* 58:1 (1999), pp. 51–84
Kean, Hilda, *Animal Rights: Political and Social Change in Britain since 1800* (London: Reaktion Books, 1998)

Li, Peter, *Animal Welfare in China* (Sydney: University of Sydney Press, 2021)

Linzey, Andrew (ed.), *The Global Guide to Animal Protection* (Urbana, University of Illinois Press, 2013)

Rothfels, Nigel, *Elephant Trails: A History of Animals and Cultures* (Baltimore: Johns Hopkins University Press, 2022)

Samanta, Samiparna, *Meat, Mercy and Morality: Animals and Humanitarianism in Colonial Bengal, 1850–1920* (Oxford: Oxford University Press, 2021)

Shevelow, Kathryn, *For the Love of Animals: The Rise of the Animal Protection Movement* (New York: Henry Holt, 2008)

Van Dooren, Thom, *Flight Ways: Life and Loss at the Edge of Extinction* (New York: Columbia University Press, 2014)

Wagschal, Steven, *Minding Animals in the Old and New Worlds: A Cognitive Historical Analysis* (Toronto: University of Toronto Press, 2018)

Index